D1196267

RADIATION EXCHANGE

An Introduction

RADIATION EXCHANGE

An Introduction

Jack H. Taylor

Department of Physics
Rhodes College
Memphis, Tennessee

ACADEMIC PRESS, INC.
Harcourt Brace Jovanovich, Publishers
Boston San Diego New York
Berkeley London Sydney
Tokyo Toronto

This book is printed on acid-free paper. ♾

ACADEMIC PRESS, INC.
1250 Sixth Avenue, San Diego, CA 92101

United Kingdom Edition published by
ACADEMIC PRESS LIMITED
24–28 Oval Road, London NW1 7DX

Library of Congress Cataloging-in-Publication Data

Taylor, Jack H.
Radiation exchange : an introduction / Jack H. Taylor.
p. cm.
ISBN 0-12-684560-3
1. Electromagnetic radiation. I. Title.
QC661.T35 1990
539.2 — dc20 89-46137
CIP

Printed in the United States of America
90 91 92 93 9 8 7 6 5 4 3 2 1

Contents

III Quantification of Electromagnetic Radiation 85

Preface

This book is about electromagnetic radiation. It deals with the radiation laws, with the phenomenon of radiation exchange and with the quantification of radiation. It can be used as a supplement to an introductory physics or introductory astronomy text and also as a guide for those members of the infrared community who would like additional insight in the area of radiation exchange. The concepts discussed herein are well within the grasp of undergraduate students.

It is quite possible that some of the concepts discussed in this book are going to be unfamiliar. For example, it might be difficult for the reader to accept the fact that he, or she, is radiating electromagnetic energy. If one is radiating, then everything is radiating and this radiation is a form of energy. What this means is that everything is radiating as well as receiving radiation and if one is going to keep up with the various net energies involved in these interactions the situation can become quite complicated. In this book the reader is shown how to come to grips with these interactions and how to demonstrate the phenomenon of radiation exchange along with many other related phenomena.

It is unfortunate that the topics discussed in this book are not covered more thoroughly in introductory physics texts than they are. The material covered herein forms the basis for the study of many exciting phenomena including such things as remote sampling, astrophysical studies, satellite studies, industrial studies, etc. The decoding of messages carried by electromagnetic radiation in remote sampling studies is a powerful investigative technique. I have always been in favor of getting undergraduates exposed to these concepts as soon as possible in their careers. These concepts open up many possibilities to them for directed inquiry projects, for honors projects and for many opportunities to involve them in meaningful experimental work in which the theory is not more than they can handle.

Having exposed undergraduates to material such as this for about three decades I can attest to the fact a meaningful research involvement is a very effective way to capture and to hold one's attention to physics. If the opportunity presents itself there is nothing like an experiment carried out on the solar atmosphere during a total solar eclipse to really fire up someone. A very important spin-off resulting from such an experience as this, or what might be called a field trip experiment, is that it forces one to become involved with the Earth's atmosphere. Electromagnetic radiation can be modified on passing through the Earth's atmosphere by the phenomena of scattering, refraction and absorption. If only we could approach the ability to observe atmospheric phenomena, and then to be able to beautifully describe the observations, as was done by the late M. Minnaert! (*The Nature of Light and Colour in the Open Air*, Dover Publications, Inc., 1954)

The style of living in the twentieth century has not been very conducive to observation of the atmosphere and skies. Many readers probably have never had the opportunity of seeing the celestial sphere down to the horizon, or, for that matter, have never seen the horizon. Many readers probably do not have enough open space where they live to be able to have an appreciation of what is meant by *daylight visual range*. Many of them have never seen the *green flash* or other setting Sun phenomena. Perhaps there are many who do not appreciate the fact that the color of the zenith twilight sky is profoundly influenced by ozone, i.e., during twilight if there were no ozone the zenith sky would become straw yellow in color as twilight progressed. The late E. O. Hulburt has pointed out that ozone, by means of its weak Chappius absorption band in the yellow, keeps the sky blue during twilight. We must all constantly force ourselves to be aware of atmospheric phenomena. In the process we are going to learn much about optical physics.

There are many problems that one needs help with in studying electromagnetic radiation that has traversed long atmospheric paths. For one thing, someone will look at a light source in the laboratory and note that it is rock stable and assume that the same will be true if they should look at a source outside the laboratory that is far away. Another problem is that most people think *visible,* i.e., they think that all electromagnetic radiation is in the visible spectrum. It is easy to understand why they think like this but it is imperative that this habit be broken, and the sooner the better.

As a teacher it is a constant battle for me to try to drive home the fact that the piece of the spectrum that one can *see* is an unbelievably small piece of the entire electromagnetic spectrum.

Another problem that most people need help with is the fact that the Earth's atmosphere is not free of absorption when one considers the infrared portion of the spectrum. For example, a person might note that someone has on a white shirt. This person will also note that when the wearer of the white shirt begins to walk away from him (or her) that the shirt continues to look white and will infer from this that the Earth's atmosphere is free of absorption because the color of the white shirt does not change as it gets farther away. This error in reasoning is brought about because one is thinking *visible.* It is imperative to appreciate the fact that if our eyes could detect infrared radiation then indeed the *color* of the white shirt would change as more and more of the light scattered off the shirt is absorbed by the Earth's atmospheric path.

This book is also concerned with the quantification of electromagnetic radiation, i.e., the measurement of wavelengths and the measurement of the intensity of the radiation (i.e., how much radiation there is.) In discussing the quantification of radiation it will be the details of measuring intensity that will be our chief concern. In addition, the chief emphasis of this book is on the infrared region of the spectrum. There are several reasons for this emphasis on the infrared. First, it is in the infrared region of the electromagnetic spectrum that so many physical phenomena can be interestingly demonstrated. Secondly, it is the spectral region where the radiation laws can be pedagogically presented. Thirdly, it is difficult to carry out quantitative measurements in the infrared and operation in this region of the spectrum has been compared to carrying out measurements in a sea of radiation where all objects are radiating and exchanging energy with each other.

Many possible research projects, other than eclipse studies, will no doubt come to mind as you read and study this book. Some that immediately come to mind are the measurement of the solar constant, the measurement of the effective radiating temperature of the ozonosphere lying above the Earth, radiometric and spectroscopic studies of lightning, spectral radiant emittance studies of the radiation from carbon arcs, radiation exchange between Earth and space and other similar experiments. It should be pointed out that radiometry offers an excellent opportunity to introduce the study of *systems analysis.*

After studying this book it is to be hoped that the reader will have developed a better feeling for the phenomenon of radiation exchange and will appreciate its importance in the scheme of things in nature.

The material contained in this book is based on a series of articles published by the Optical Society Of America in *Applied Optics* and *Optics News.* The article in *Applied Optics* was published in February, 1987, and the articles in *Optics News* were published approximately monthly from January, 1987, to December, 1987. The Optical Society has graciously granted permission for the republication of the material in its present form.

Part I

Key Ideas of Radiation Exchange

Chapter 1

Radiation Exchange

1.1 Introduction

It has been my observation that most texts provide a minimal coverage of the phenomenon of radiation exchange. What little material there is on radiation exchange in most introductory texts is usually covered in the section dealing with heat, in particular, in the section dealing with the transfer of heat by radiation. In fact, radiation exchange is not listed in the indices of the most frequently used texts for physics majors in this country. The author has found this to be an area of considerable interest to most people.

It is important to have some understanding of the key ideas involved in radiation exchange. Understanding and appreciating radiation exchange is necessary for effective work in infrared radiometry. Radiation exchange is an area of physics where one can use, and needs, as much help from the textbook as possible. Phenomena involving radiation exchange are everywhere; one simply needs help in recognizing them. The subject of radiation exchange, being extremely open-ended, holds great possibilities for the reader. You will find that it is laden with astrophysical and space applications, both of which interest most people. Depending on the extent to which one develops this area, it has been my experience that many ideas and suggestions for student experiments, directed inquiries, and honors projects are quite likely to arise. In this connection, let me say that I routinely have my students measure the effective radiating temperature of the ozonosphere above Memphis.

1.2 Everything Is Radiating

1.2.1 Stefan-Boltzmann Law

A good place to start a discussion of radiation exchange is with the Stefan-Boltzmann Law. Most texts write the law as

$$H = A\epsilon\sigma T^4 \tag{1.1}$$

and state that this gives the rate of radiation of energy from a surface of area A and absolute temperature T and that ϵ (the radiant emissivity) is a dimensionless number between 0 and 1. The quantity σ is the Stefan-Boltzmann constant. This law was first arrived at empirically. The theoretical basis on which it rests is the Plank Radiation Law, which is discussed in Chapter 3.

Solving the above equation for H/A one obtains the units of watts/area. This is where most texts leave the units of Equation (1.1). They fail to point out that the units of H/A are really watts/area radiated into a hemisphere. This failure to emphasize that this is indeed radiation into a hemisphere can be quite frustrating to someone who is trying to calculate the power from a radiating object that is intercepted by a mirror, using only what most introductory texts state as being the units of Equation (1.1).

Throughout this book, we shall be talking about solid angles (i.e., about steradians). Figure 1.1 is a sketch showing the definition of both a plane angle and a solid angle. Since $d\Omega = dA/R^2$, one determines the solid angle surrounding a point, Ω, by integrating over the entire sphere to obtain $4\pi R^2/R^2 = 4\pi$. In other

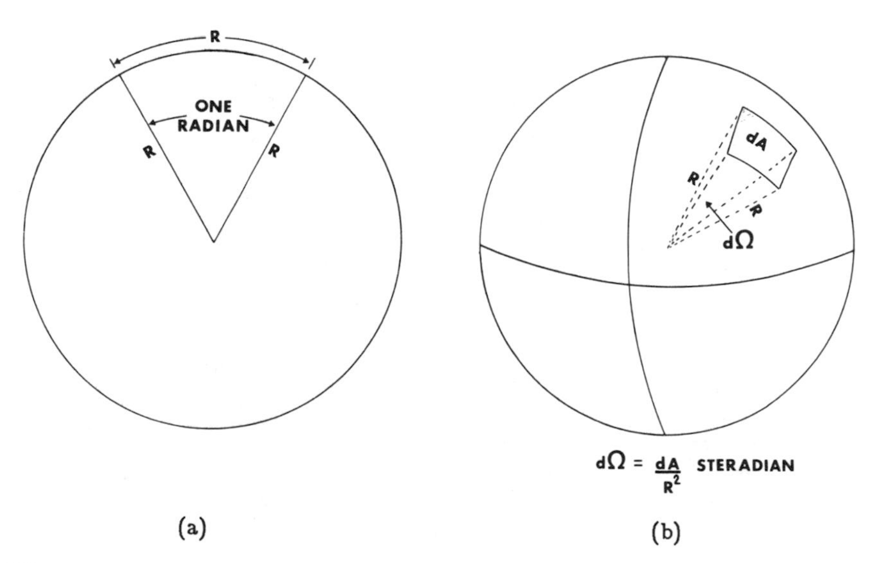

Figure 1.1: A review of angles. (a) A two-dimensional (or plane) angle. (b) A three-dimensional (or solid) angle. ***Steradian*** comes from the Greek word *stereós*, meaning solid.

words, there are 4π steradians surrounding a point.

Since we are talking about radiation into a hemisphere, one might easily reason that $H/A/2\pi$ would have the units of watts/area/steradian. The units are correct, but we should have divided by π and not 2π for reasons that are to be discussed shortly.

In present-day radiometric terminology a quantity with the units of watts/unit area/steradian is called the *radiance* and is a useful concept in radiometry.

Equation (1.1) contains many of the key ideas of radiation exchange. It clearly shows that all objects that have a temperature radiate, and since one cannot achieve absolute zero, it means that all objects are radiating. An understanding of this important equation will come with continued use and study.

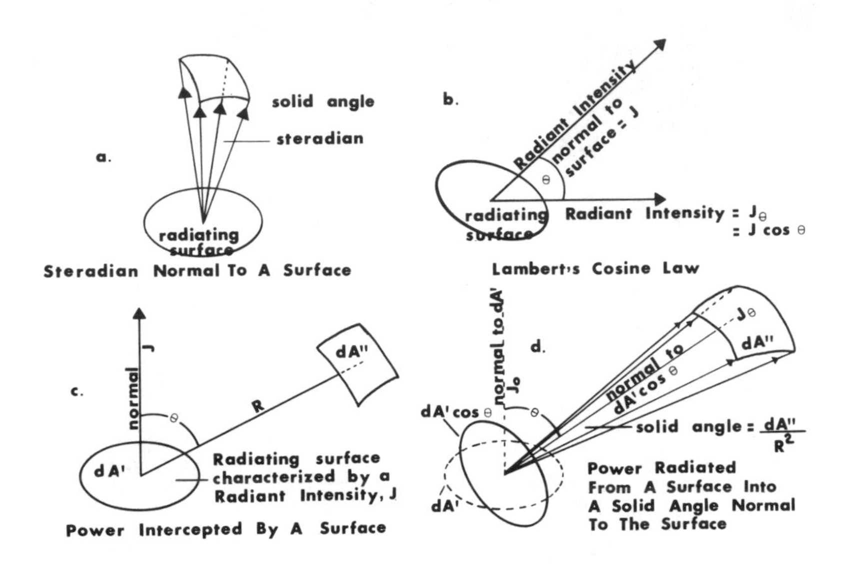

Figure 1.2: Development of the concept of radiance.

1.2.2 Development of the Concept of Radiance

Suppose we are interested in the amount of power radiated per unit area into a steradian normal to the radiating surface as shown in Figure 1.2(a).

The idea of a point source radiating uniformly in all directions is a useful concept theoretically but one that is difficult to achieve in practice. For the majority of cases encountered, the source will be what is termed an *extended source*, that is, it will have a measurable area and must be treated differently than a point source. For a point source we could characterize it quite adequately by stating the number of watts per steradian radiated in a given direction. For an extended source we could characterize it by stating the number of watts per unit area per steradian radiated normal to the surface.

As we have noted, the term *radiance* is used to describe the power per unit solid angle per unit area of source radiated by the source. Many extended sources obey what is known as Lambert's Cosine Law of Radiation. Such sources are said to be Lambertian or perfectly diffuse. If the radiant intensity (defined as watts/steradians) of an extended source obeys Lambert's Cosine Law, then for all wavelength intervals the radiant intensity at some angle θ with respect to the normal to the surface varies as the cosine of the angle [see Figure 1.2(b)], and we can write

$$J_\theta = J \cos\theta \quad \text{Lambert's Cosine Law of Radiation.} \tag{1.2}$$

In order to realize Equation (1.2) experimentally, it is necessary that the point of observation be far enough away that the radiating surface can be considered a point source.

Consider the extended source shown in Figure 1.2(c), which we assume obeys Lambert's Cosine Law of Radiation. Let us now calculate the power intercepted by the element of area dA'' located a distance R away. The radiant intensity in the direction of dA'' is given by $J\cos\theta$. The size of the radiating surface that is normal to R is $dA'\cos\theta$. The ratio of $J\cos\theta$ and the projected area normal to R is given by the following:

$$\frac{J\cos\theta}{dA'\cos\theta} = \frac{J}{dA'} = \text{constant, independent of the angle } \theta. \tag{1.3}$$

The quantity J/dA' has the units of watts/steradian/area and is what we are calling the radiance. In other words, for a surface obeying Lambert's Cosine Law of Radiation, the radiance is a constant independent of θ. The reason for this is that the emitted radiation per steradian falls off with $\cos\theta$ and the "projected" surface area of the radiating source falls off at the same rate.

The term *radiance* is used primarily for work in the infrared. For work in the visible the quantity corresponding to radiance is *brightness*, and we can restate Lambert's Cosine Law of radiation by saying that the brightness of a blackbody or of a diffuse source is the same, regardless of the angle from which it is viewed. This law does not hold strictly true for many non-blackbody substances. The law is nearly true for matt white surfaces (for example, those composed of magnesium oxide and plaster of Paris). It is because of this law that incandescent spherical and cylindrical bodies appear to the eye as flat disks or ribbons. It almost applies in the case of the Sun but not quite. Because of the radial density distribution of luminous mass, the Sun does not emit diffusely, and accurate photometric measurements show the existence of a darker edge.

Let us now calculate the power intercepted by the area dA'' shown in Figure 1.2(d):

$$J_\theta = J_0 \cos\theta; \tag{1.4}$$

$$J_0 = N\ dA', \text{ where } N = \text{radiance}. \tag{1.5}$$

Power intercepted by $dA'' = P$:

$$P = [J_\theta]\frac{\text{watts}}{\text{sr}}\left[\frac{dA''}{R^2}\right]\text{sr} \tag{1.6}$$

$$= [N]\frac{\text{watts}(dA'\cos\theta)}{(\text{area})(\text{sr})}\text{area}\left[\frac{dA''}{R^2}\right]\text{sr}. \tag{1.7}$$

Note: $[dA'\ \cos\theta]$ is the projected area seen by dA'' (i.e., normal to R).

In Figure 1.3, let the area of the radiating surface be A and its radiance be N. Our problem is to calculate the power intercepted by the shaded annulus and then to sum over all annuli on the hemispherical surface:

$$\begin{aligned}\text{Area of shaded annulus} &= (2\pi R\sin\theta)(Rd\theta)\\ &= 2\pi R^2\sin\theta d\theta.\end{aligned} \tag{1.8}$$

Let dP be the power intercepted by the shaded annulus:

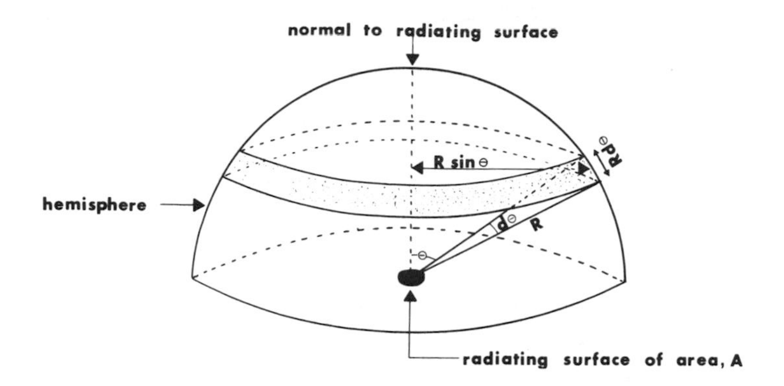

Figure 1.3: Power intercepted by an annulus.

$$
\begin{aligned}
dP &= [N\cos\theta]\frac{\text{watts}}{(\text{area})(\text{sr})}[A]\text{area}\Big[\frac{2\pi R^2\sin\theta d\theta}{R^2}\Big]\text{sr}, \\
dP &= 2\pi NA\sin\theta\cos\theta d\theta \text{ watts}; \qquad (1.9) \\
P &= \text{total power radiated into hemisphere}, \\
P &= \int_0^P dP = \int_0^{\frac{\pi}{2}} 2\pi NA\sin\theta\cos\theta d\theta, \\
P &= 2\pi NA\int_0^{\frac{\pi}{2}}\sin\theta\cos\theta d\theta = 2\pi N\,A\Big[\frac{\sin^2\theta}{2}\Big]_0^{\frac{\pi}{2}}, \\
P &= \pi NA \text{ watts}. \qquad (1.10)
\end{aligned}
$$

From this result, we see that if P is the power in watts radiated into a hemisphere, the number of watts per unit area per steradian radiated by the source is given by $P/(\pi A)$, which is called the radiance N.

There are two factors responsible for the factor of π instead of 2π, namely,

1. The constancy of the radiance N with viewing angle θ,

2. The fact that the projected area of the radiating surface normal to the viewing angle depends on the cosine of the viewing angle.

1.2.3 Demonstrating That Everything Is Hot

Statement

The phenomenon of radiation exchange is largely responsible for the poor precision attainable in infrared radiometry and infrared spectroscopy. Carrying out experiments in the infrared is analogous to a kind of photometry in which all the equipment is red hot and even the air of the laboratory is red hot. The fact that everything is hot, so to speak, can be demonstrated initially with very simple and modest equipment. After the reader has developed a feeling for the subject, more elegant apparatus will be used.

Figure 1.4 is a sketch of the apparatus used throughout this book to demonstrate the key ideas of radiation exchange. Perhaps a few comments on the radiation thermopile and spotlight galvanometer would be appropriate at this time. The Leslie cube and Crookes' radiometer are discussed later in the book. A radiation thermopile consists of several radiation thermocouples connected in series. A radiation thermocouple consists of two junctions of dissimilar metals. If one of the junctions is at a different temperature than the other, a difference of potential is developed (i.e., a voltage is generated). If an ammeter is connected between the two junctions, a current will flow. In the apparatus sketched in Figure 1.4, the spotlight galvanometer is used to measure the current. The greater the difference in the temperature between the hot and cold junctions of the thermopile, the greater will be the deflection of the spotlight galvanometer. In the construction of the thermopile the "hot" junctions are located (i.e., thermally connected) behind a blackened piece of very thin metal

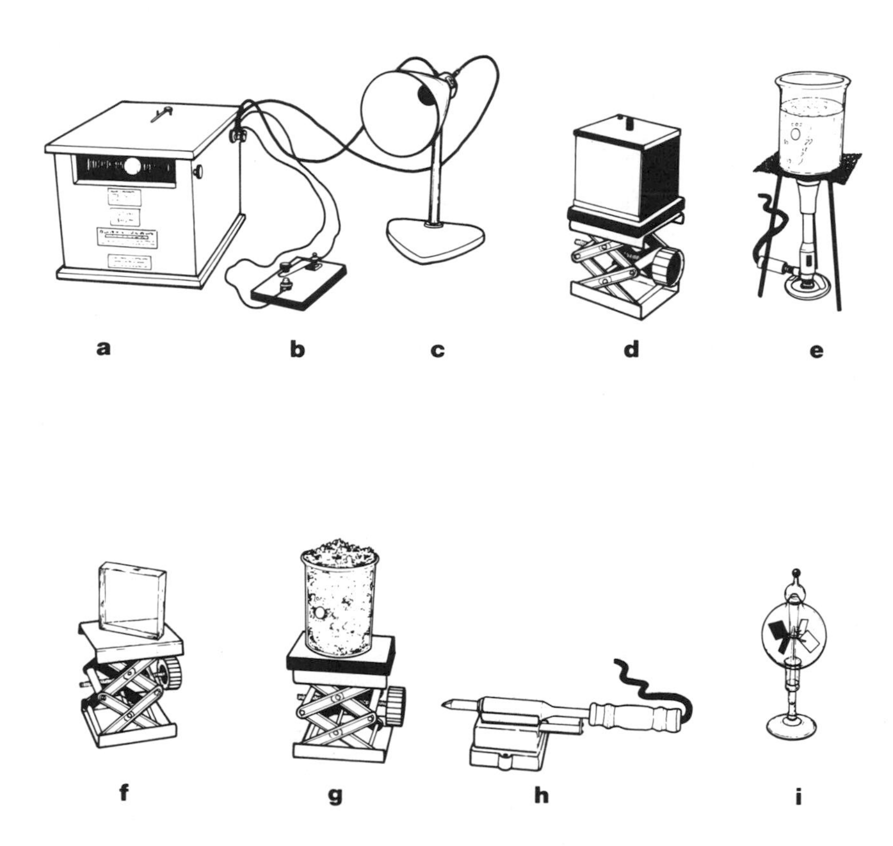

Figure 1.4: Apparatus used by the author to demonstrate radiation exchange: a = spotlight galvanometer, b = damping key, c = radiation thermopile, d = Leslie cube, e = beaker of hot water (used to fill the Leslie cube), f = thick piece of high-quality glass, g = beaker of ice, h = soldering iron, and i = Crookes' radiometer.

that is exposed to the radiation to be detected, while the "cold" junctions are protected (i.e., shielded) from the radiation that is being detected. The damping key shown in Figures 1.4 and 1.5 is used to put an electrical short circuit across the input to the spotlight galvanometer in case the input causes the spotlight to move off scale. Since the impedance of the thermopile is low (i.e., about 25 ohms), it is important that the spotlight galvanometer used has a comparable impedance.

I usually use a 200 watt soldering iron, and when the iron is energized I find that it must be placed several feet in front of the thermopile to keep the galvanometer deflection on scale. In the case of the beaker of ice, I support the beaker on a lab jack and place it immediately in front of the reflecting cone of the thermopile.

Students often find it difficult to appreciate the fact that the beaker of ice is also radiating, since they certainly cannot detect this radiation with their eyes! In this demonstration the instructor starts with the galvanometer zeroed at about mid-scale. Radiation from the hot soldering iron will produce a large deflection, to the left or right of zero, depending on how the galvanometer and thermopile have been connected. Radiation from the beaker of ice produces a deflection in a direction opposite to that produced by the soldering iron.

Figure 1.5 is a schematic diagram illustrating the demonstration of radiation exchange.

The students rapidly begin to appreciate the fact that the visible portion of the spectrum indeed involves only a very tiny portion of the entire electromagnetic spectrum and that the optimum region of the spectrum for the majority of the radiation exchange demonstrations lies in the infrared.

As indicated in Figure 1.5, the importance of blackbodies plays a very significant role in this area of radiation exchange. Blackbodies are important because they are good absorbers of radiation. However, it should be pointed out that blackbodies do not necessarily look black; it all depends on their tempera-

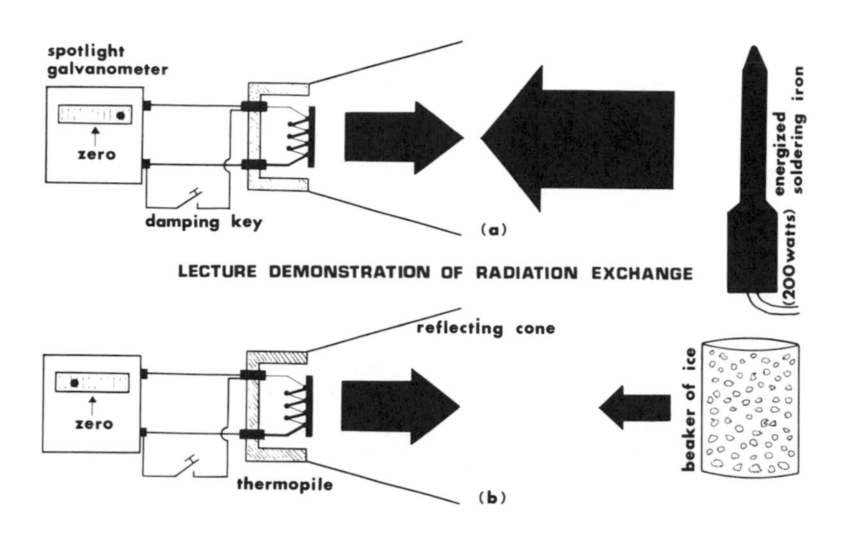

Figure 1.5: Schematic diagram illustrating the demonstration of radiation exchange.

ture. If a blackbody has a temperature above ~500° C, it appears incandescent (dull glow only).

In Figure 1.5 the size of the blackened arrows is proportional to the radiance. As a result, we see that in Figure 1.5(a) the net radiation is *in* toward the thermopile, while in Figure 1.5(b) the net radiation is *out* toward the ice. From a simple lecture demonstration such as this one, it is easy to see how such a technique can become a powerful remote sampling technique; by simply pointing the radiometer at an object, one can tell by noting the direction of deflection of the spot on the spotlight galvanometer whether the object is hotter or colder than the thermopile. In Figure 1.5 it should be emphasized that the size of the arrrow near the thermopile depends on the temperature of the thermopile and that since the thermopile has a temperature it is radiating.

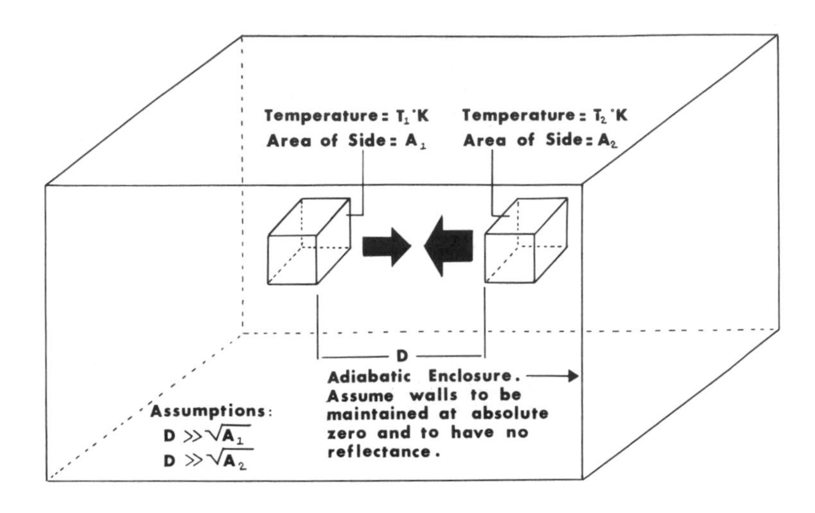

Figure 1.6: Radiation exchange between two objects in an adiabatic enclosure.

Explanation of our demonstration

Figure 1.6 is included to assist the reader in understanding why a hot soldering iron located in front of a radiation thermopile will produce a deflection in a certain direction on the spotlight galvanometer and a beaker of ice will produce a deflection in the opposite direction. Assume the two objects inside the adiabatic enclosure are blackbodies, i.e., they reflect none of the radiation falling on them. (Blackbody radiation is discussed in Chapter 5.) This assumption is equivalent to stating that the radiant emissivity of each object is equal to unity (i.e., in Equation (1.1) ϵ has the value of unity). In addition, let us assume that the distance between the two objects is many times greater than the linear dimensions of either object. This assumption simplifies the geometrical considerations without in any way detracting from the pedagogy of the presentation. We assume that only the two facing sides of the objects can see each other and, therefore, are the only sides involved in considerations of radiation exchange. As a further postulate, let us assume that the radiant reflectance of the walls of the adiabatic enclosure is zero. As a result of all these assumptions, we have only two surfaces with which to contend as we investigate the radiation exchange.

In this example, $T_1 \neq T_2$ and $A_1 \neq A_2$. Let us suppose that A_1 represents a thermal detector (say a thermocouple, bolometer, or Golay cell) and A_2 represents some other object. We shall now write an expression for the power striking A_1 and A_2. Let

$$\begin{aligned} P_2 &= \text{power absorbed by } A_1 \\ &= \frac{\sigma T_2^4}{\pi} A_2\, A_1/D^2 \text{ watts} \end{aligned} \tag{1.11}$$

(i.e., A_1 sees only A_2);

$$\begin{aligned} P_1 &= \text{power absorbed by } A_2 \\ &= \frac{\sigma T_1^4}{\pi} A_1\, A_2/D^2 \text{ watts} \end{aligned} \tag{1.12}$$

(i.e., A_2 sees only A_1).

Consider next A_1, which we have assumed to be a thermal detector. The output (or voltage developed) of A_1 will depend on its temperature change which is proportional to the net power striking A_1:

$$\begin{aligned} \Delta P &= \text{net power absorbed by} A_1 = P_2 - P_1 \\ &= \text{response of thermal detector,} \\ \Delta P &= \frac{\sigma T_2^4}{\pi} A_2 A_1/D^2 - \frac{\sigma T_1^4}{\pi} A_1 A_2/D^2, \\ \Delta P &= \frac{\sigma A_1 A_2}{\pi D^2}[T_2^4 - T_1^4]. \end{aligned} \tag{1.13}$$

This result for ΔP readily explains the results of the lecture demonstration on radiation exchange previously described, namely, the response is positive if $T_2 > T_1$, negative if $T_2 < T_1$, and there is no response if $T_2 = T_1$.

Let us next consider the case where A_1 is a blackbody and A_2 is no longer a blackbody. Now assume A_2 to be what is referred to as a *graybody*, i.e., its radiant emissivity is less than unity and has the same value at all wavelengths. Let this value of the radiant emissivity be ϵ_2. From analogy with what we previously did, we can write

$$\begin{aligned}
P_2' &= \text{power absorbed by} A_1 \\
&= \frac{\epsilon_2 \sigma T_2^4}{\pi} A_2 A_1 / D^2 \text{ watts;} \qquad (1.14) \\
P_1' &= \text{power absorbed by} A_2 \\
&= \frac{\sigma T_1^4}{\pi} A_1 A_2 / D^2 \text{ watts;} \qquad (1.15) \\
\Delta P' &= \text{net power absorbed by} A_1 = P_2' - P_1' \\
&= \frac{\sigma A_1 A_2}{\pi D^2} [\epsilon_2 T_2^4 - T_1^4] \text{ watts.} \qquad (1.16)
\end{aligned}$$

This expression for $\Delta P'$ requires careful examination. Assuming the validity of the second law of thermodynamics, the expression for $\Delta P'$ is wrong, because it predicts a net power transfer and hence a net energy transfer between two bodies at the same temperature without the performance of any work. We must have made a mistake!

The explanation is readily forthcoming. The expression for P_1' is incorrect. The correct expression is

$$P_1' = \frac{\epsilon_2 \sigma T_1^4}{\pi} A_1 A_2 / D^2 \text{ watts,} \qquad (1.17)$$

since from Kirchhoff's Radiation Law (discussed in Chapter 4) the radiant emissivity is numerically equal to the radiant absorptance. Using the correct value for P_1' we write

$$\begin{aligned}\Delta P_1' &= \text{net power absorbed by} A_1 = P_2' - P_1' \\ &= \frac{\epsilon_2 \sigma A_1 A_2}{\pi D^2}[T_2^4 - T_1^4] \text{ watts.} \qquad (1.18)\end{aligned}$$

This correct value for $\Delta P'$ can readily be extended to the case where both the detector and object are graybodies. Let the radiant emissivity of A_1 be ϵ_1 and of A_2 be ϵ_2:

$$\begin{aligned}P_2'' &= \text{power absorbed by} A_1 \\ &= \frac{\epsilon_1 \epsilon_2 \sigma}{\pi} T_2^4 A_2 A_1 / D^2 \text{ watts;} \qquad (1.19) \\ P_1'' &= \text{power absorbed by} A_2 \\ &= \frac{\epsilon_1 \epsilon_2 \sigma}{\pi} T_1^4 A_1 A_2 / D^2 \text{ watts;} \qquad (1.20) \\ \Delta P'' &= \text{net power absorbed by} A_1 = P_2'' - P_1'' \\ &= \frac{\epsilon_1 \epsilon_2 \sigma A_1 A_2}{\pi D^2}[T_2^4 - T_1^4] \text{ watts.} \qquad (1.21)\end{aligned}$$

In any practical application, the calculation of the response of a thermal detector is considerably more complicated than this highly hypothetical case of radiation exchange between only two objects; i.e., it would differ in degree but not in kind.

1.3 Building a Radiometer

1.3.1 Field of View of an Optical System

Let us now consider the challenge of locating a thermal detector in the focal plane of a lens or mirror (i.e., we want to make a radiometer), viewing a distant object and predicting the response of the detector. In order to carry out this calculation it will be necessary to consider the concept of field of view of an optical system. After considering the problem for a while, one will realize that one of two possibilities will arise:

1. The field of view of the optical system is smaller than the object to be studied.

2. The field of view of the optical system is larger than the object to be studied.

Figures 1.7-1.10 have been included to assist in these computations. In these figures a refracting radiometer (i.e., the use of a lens) is shown, whereas in the majority of cases one would choose to use a paraboloidal front-surface mirror instead. A simple lens is shown for pedagogical purposes.

In these calculations we want to consider a distant object such as a cloud, mountain, ocean surface, desert floor, or planet. Obviously, in this application the atmospheric path is considerably long. To simplify matters at this stage, we assume there is no absorption and no scattering in the atmosphere. We shall assume that the distance between lens and object, or background, is many times either of the linear dimensions of the detector and that there is neither absorption nor reflection loss at the lens.

1.3.2 Field of View Smaller Than Object Being Studied

Consider Figure 1.7 and calculate the power absorbed by the detector. Let

$P_2 =$ power from the object under study intercepted by the lens (and that is absorbed by the detector)

$$= \frac{\sigma T_2^4}{\pi}\Big[\text{area of object seen by the detector}\Big]\Big[\frac{\text{area of lens}}{D_2^2}\Big]. \tag{1.22}$$

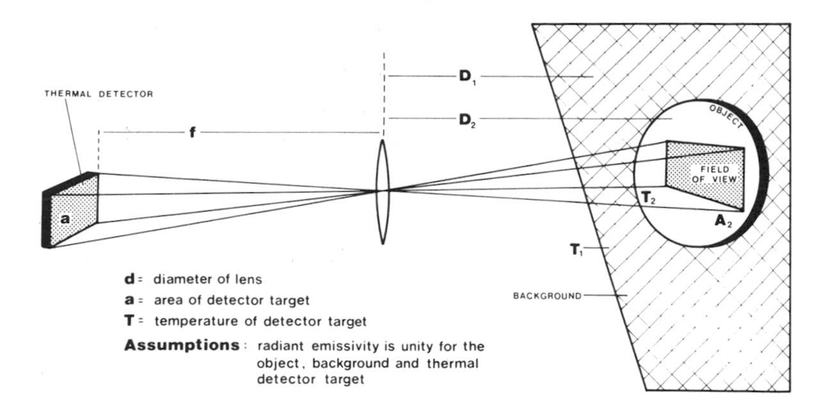

Figure 1.7: Radiometer with a field of view smaller than the object and viewing the object.

The area of the object that is seen by the detector is the field of view and is given by

$$\left\{a\left[\frac{D_2}{f}\right]^2\right\}.$$

The quantity

$$\left[\frac{\text{area of lens}}{D_2^2}\right]$$

is the solid angle in steradians subtended by the lens at the object:

$$\begin{aligned} P_2 &= \frac{\sigma T_2^4}{\pi} a[D_2/f]^2 \frac{\pi d^2}{4D_2^2} \\ &= \frac{\sigma T_2^4 a d^2}{4f^2} \text{ watts}, \end{aligned} \tag{1.23}$$

where a is the area of the detector target, d is the diameter of the lens, and f is the focal length.

Many people are surprised by this expression for P_2, namely, that the power incident on the detector is independent of the distance to the object. They readily accept it when it is pointed out that the field of view increases as D_2^2 and the solid angle subtended by the lens at the object decreases as $1/D_2^2$.

The ratio f/d (i.e., the focal length of the lens divided by the diameter of the lens) is defined as the *F/number* of the lens or mirror and is written as the *F/no.*. Using this definition of the F/no. we can express P_2 as

$$P_2 = \left[\frac{\sigma T_2^4 a}{4(F/\text{no.})^2}\right]\text{watts.} \tag{1.24}$$

Writing P_2 in this form makes explicit the following points:

1. The power incident on the detector varies inversely as the F/no. squared, i.e., an $F/2$ lens collects four times as much power as an F/4 lens.

2. The power intercepted by a large diameter lens is not necessarily greater than that intercepted by a smaller diameter lens.

Let P = power radiated by the detector itself. By this we mean the power radiated by the detector that is intercepted by the lens:

$$P = \frac{\sigma T^4}{\pi} a \Big[\text{solid angle subtended by the lens at the detector}\Big] \text{ watts,} \tag{1.25}$$

$$P = \frac{\sigma T^4}{\pi} a \left[\frac{\pi d^2}{4f^2}\right] = \frac{\sigma T^4 a d^2}{4f^2} \text{ watts.} \tag{1.26}$$

The net power striking the detector when it sees only the object is given by the following expression:

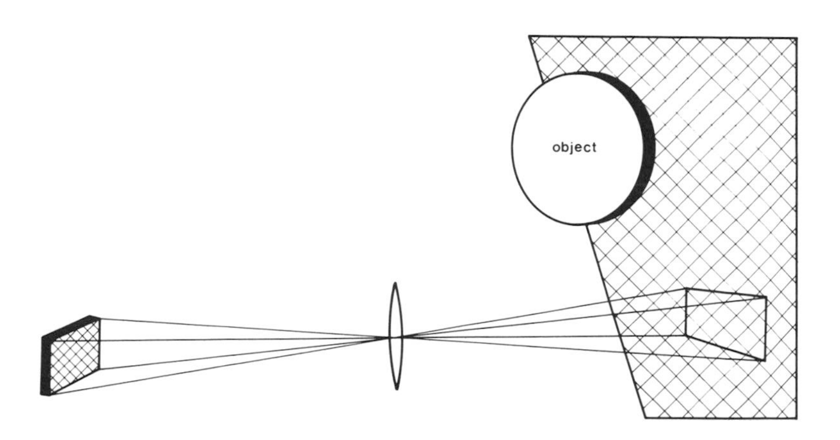

Figure 1.8: Radiometer with a field of view smaller than the object and viewing the background.

$$\Delta P_{\text{Figure 1.7}} = P_2 - P$$
$$= \left[\frac{\sigma T_2^4 a d^2}{4f^2}\right] - \left[\frac{\sigma T^4 a d^2}{4f^2}\right], \quad (1.27)$$
$$\Delta P_{\text{Figure 1.7}} = \frac{\sigma a d^2}{4f^2}[T_2^4 - T^4] \text{ watts.} \quad (1.28)$$

Suppose we now point the radiometer so the object under study is no longer in the field of view and the detector sees only the background as indicated in Figure 1.8. Let

P_1 = power from background intercepted by the lens (and that is absorbed by the detector),

$$P_1 = \frac{\sigma T_1^4}{\pi} a [D_1/f]^2 \frac{\pi d^2}{4D_1^2}, \quad (1.29)$$
$$P_1 = \frac{\sigma T_1^4 a d}{4f^2} \text{ watts.} \quad (1.30)$$

Let P = power radiated by the detector. Then

$$P = \frac{\sigma T^4}{\pi} a \frac{\pi d^2}{4f^2} = \frac{\sigma T^4 a d^2}{4f^2} \text{ watts.} \tag{1.31}$$

The net power striking the detector when it sees only the background is

$$\begin{aligned} \Delta P_{\text{Figure 1.8}} &= P_1 - P \\ &= \left[\frac{\sigma T_1^4 a d^2}{4f^2}\right] - \left[\frac{\sigma T^4 a d^2}{4f^2}\right], \end{aligned} \tag{1.32}$$

$$\Delta P_{\text{Figure 1.8}} = \frac{\sigma a d^2}{4f^2}[T_1^4 - T^4] \text{ watts.} \tag{1.33}$$

The net response, or net power striking the detector, is

$$\Delta P = \Delta P_{\text{Figure 1.7}} - \Delta P_{\text{Figure 1.8}},$$

$$\Delta P = \left[\frac{\sigma T_2^4 a d^2}{4f^2} - \frac{\sigma T^4 a d^2}{4f^2}\right] - \left[\frac{\sigma T_1^4 a d^2}{4f^2} - \frac{\sigma T^4 a d^2}{4f^2}\right], \tag{1.34}$$

$$\Delta P = \frac{\sigma a d^2}{4f^2}[T_2^4 - T_1^4] \text{ watts.} \tag{1.35}$$

This expression for ΔP (i.e., the case in which the field of view is completely filled by either the object or the background) predicts the following:

1. The response of the detector is independent of the distance to either the object or the background.
2. The response of the detector varies inversely with the square of the F/no. of the optical system.
3. The response of the detector varies directly as the area of the detector.
4. The response of the detector varies directly with the field of view expressed in steradians. It will be noted that (a/f^2) is the field of view in steradians.

5. The response of the detector is independent of the temperature of the detector.

There are many interesting practical applications in which T_1 and T_2 will be nearly equal. It is interesting to look at the expression for ΔP under the assumption that $T_2 \approx T_1$:

$$\begin{aligned} T_2^4 - T_1^4 &= (T_2^2 + T_1^2)[(T_2^2 - T_1^2)] \\ &= (T_2^2 + T_1^2)[(T_2 + T_1)(T_2 - T_1)]. \end{aligned} \quad (1.36)$$

However,

$$\begin{aligned} T_2^2 + T_1^2 &\approx 2T_1^2; \\ T_2 + T_1 &\approx 2T_1; \\ T_2 - T_1 &= \Delta T; \\ (T_2^4 - T_1^4) &\approx 4T_1^3 \Delta T. \end{aligned}$$

Substituting the latter value for $(T_2^4 - T_1^4)$ into the expression for ΔP, one obtains

$$\Delta P = \frac{\sigma a d^2}{4 f^2}(4T_1^3 \Delta T) \quad (1.37)$$

or

$$\Delta P = \frac{\sigma a d^2}{f^2} T_1^3 \Delta T. \quad (1.38)$$

Thus, for a given ΔT between object and background, the response of the detector varies directly as T_1^3.

1.3.3 Field of View Larger Than Object Being Studied

Consider next the situation shown in Figure 1.9. In this case, the field of view is larger than the object, or, to put it another way,

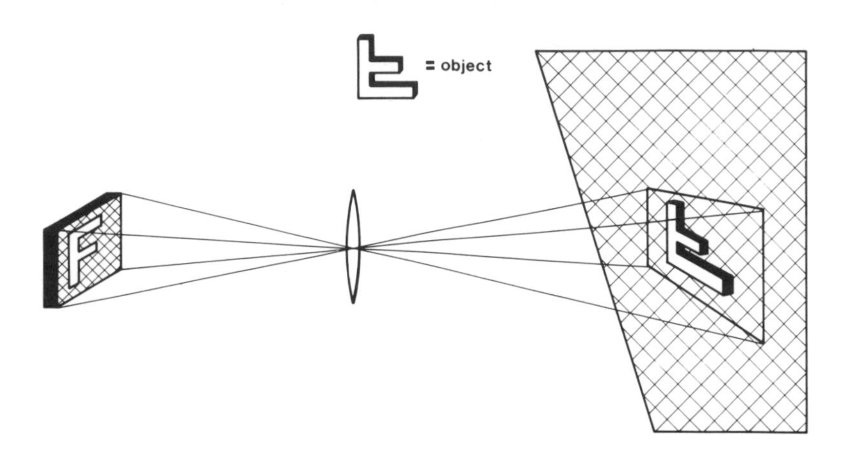

Figure 1.9: Radiometer with a field of view larger than the object and viewing the object plus part of the background.

the image of the object formed on the detector is not large enough to fill completely the face of the detector. A good pedagogical approach to this case is to ask the following questions:

1. What is the area of the detector covered by the image of the object under study?

2. What is the area of the detector covered by the image of the background?

The area of the image of the object A_2 formed on the detector is given by $A_2(f/D_2)^2$.

The area of the detector covered by the image of the background equals the area of the detector available for seeing the background and is given by the following expression: $[a - A_2(f/D_2)^2]$. The field of view of that part of the area of the detector that is available for seeing the background is given by the following expression: $[a - A_2(f/D_2)^2](D_1/f)^2$. Let

$P_2' =$ power from the entire side of the object facing the detector plus that part of the background that is not obscured by the object. Note: This is the power intercepted by the lens and is seen by the detector.

$$P_2' = \frac{\sigma T_2^4}{\pi} A_2 \frac{\pi d^2}{4D_2^2} + \frac{\sigma T_1^4}{\pi}[a - A_2(f/D_2)^2] \times (D_1/f)^2 \left[\frac{\pi d^2}{4D_1^2}\right] \text{ watts.} \tag{1.39}$$

Let

$$\begin{aligned} P' &= \text{power radiated by the detector} \\ &= \frac{\sigma T^4}{\pi} a \frac{\pi d^2}{4f^2} = \frac{\sigma T^4 a d^2}{4f^2} \text{ watts;} \qquad (1.40) \\ \Delta P'_{\text{Figure 1.9}} &= P_2' - P' \\ &= \frac{\sigma T_2^4}{\pi} A_2 \frac{\pi d^2}{4D^2} + \frac{\sigma T_1^4}{\pi} a \frac{\pi d^2}{4D_1^2}(D_1/f)^2 \\ &\quad - \frac{\sigma T_1^4}{\pi} A_2 (f/D_2)^2 \frac{\pi d^2}{4D_1^2}(D_1/f)^2 \\ &\quad - \frac{\sigma T^4 a d^2}{4f^2}, \qquad (1.41) \\ \Delta P'_{\text{Figure 1.9}} &= \frac{\sigma T_2^4}{4D_2^2} + \frac{\sigma T_1^4 a d^2}{4f^2} - \frac{\sigma T_1^4 A_2 d^2}{4D_2^2} \\ &\quad - \frac{\sigma T^4 a d^2}{4f^2} \text{ watts.} \qquad (1.42) \end{aligned}$$

We now point the radiometer so that the detector sees only the background, as shown in Figure 1.10. Let

$P_1' =$ power from background intercepted by lens (and that is absorbed by the detector),

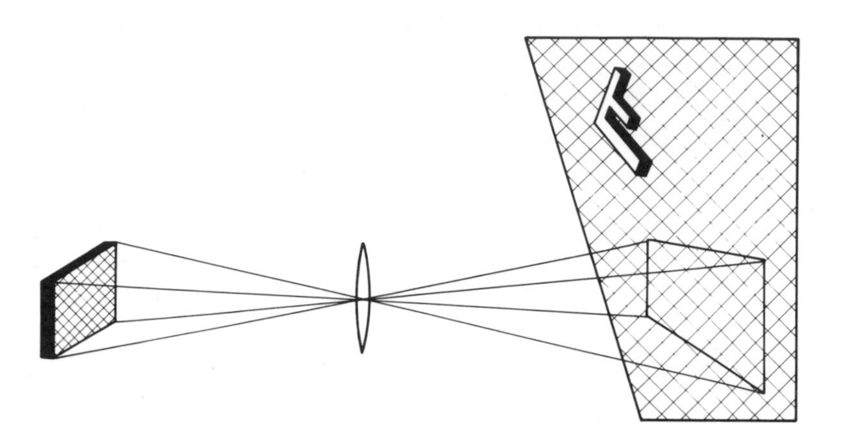

Figure 1.10: Radiometer with a field of view larger than the object and viewing the background.

$$P_1' = \frac{\sigma T_1^4}{\pi} a (D_1/f)^2 \frac{\pi d^2}{4D_1^2},$$

$$P_1' = \frac{\sigma T_1^4 a d^2}{4f^2} \text{ watts.} \tag{1.43}$$

Let

P' = power radiated by the detector, i.e., the power radiated by the detector that is intercepted by the lens,

$$P' = \frac{\sigma T^4}{\pi} a \frac{\pi d^2}{4f^2} = \frac{\sigma T^4 a d^2}{4f^2} \text{ watts;} \tag{1.44}$$

$$\Delta P'_{\text{Figure 1.10}} = P_1' - P'$$

$$= \frac{\sigma T_1^4 a d^2}{4f^2} - \frac{\sigma T^4 a d^2}{4f^2}. \tag{1.45}$$

The net response, or net power striking the detector, is

$$\begin{aligned}
\Delta P' &= \Delta P'_{\text{Figure 1.9}} - \Delta P'_{\text{Figure 1.10}} \\
&= \frac{\sigma T_2^4 A_2 d^2}{4D_2^2} + \frac{\sigma T_1^4 a d^2}{4f^2} - \frac{\sigma T_1^4 A_2 d^2}{4D_2^2} \\
&\quad - \frac{\sigma T^4 a d^2}{4f^2} - \frac{\sigma T_1^4 a d^2}{4f^2} + \frac{\sigma T^4 a d^2}{4f^2}, \qquad (1.46)
\end{aligned}$$

$$\Delta P' = \frac{\sigma T_2^4 A_2 d^2}{4D_2^2} - \frac{\sigma T_1^4 A_2 d^2}{4D_2^2}, \qquad (1.47)$$

$$\Delta P' = \frac{\sigma A_2 d^2}{4D_2^2}[T_2^4 - T_1^4] \text{ watts.} \qquad (1.48)$$

It is interesting to note that for this case, i.e., for the field of view larger than the object, the net response of the detector

1. depends on the area of the object under study,
2. varies inversely as the square of the distance to the object,
3. is independent of the temperature of the detector,
4. is independent of the area of the detector,
5. is proportional to the area of the lens (or mirror),
6. is independent of the F/no. of the optical system.

After this brief quantitative exercise the reader should have some feeling for radiation exchange. In addition, the readers who decide to do a directed inquiry or an honors project in optical physics should have some feeling for how to proceed. For example, the readers who decide to undertake a study of planetary radiometry should now realize that the size of the collecting mirror determines the amount of power reaching the detector. By the same token, the readers who undertake a radiometric study from a helicopter looking down at the ocean (or desert) should realize that the F/no. of the mirror determines the amount of power striking the detector and not the area of the mirror. Continuing along these lines the reader might consider what type of optical system would be most appropriate for studying radiometrically the radiation from lightning and from meteors.

One of the best articles dealing with radiation exchange, as well as with other infrared radiometric subjects, of which I am aware appeared in *Guidance (Principles of Guided Missile Design)*, Edited by Grayson Merrill, in particular, in Chapter 5, "Emission, Transmission, and Detection of The Infrared" by John A. Sanderson (Van Nostrand, New York, 1955).

Chapter 2

Energy Associated with Electromagnetic Radiation

In Chapter 1 it was pointed out that a key idea in radiation exchange is that everything is radiating. In this chapter we are going to discuss the energy associated with electromagnetic radiation, and we are going to describe two lecture demonstrations of this phenomenon.

2.1 Tissue Paper and Flashbulb Demonstration

Perhaps some readers will have had an experience from scouting in which they used a small lens to focus solar radiation and start the burning of dry leaves. The physical interpretation of that experience, however, is sometimes forgotten.

I like to demonstrate that energy is associated with electromagnetic radiation using the apparatus sketched in Figure 2.1. In this demonstration, I use a large clear flashbulb located in the focal plane of a parabolic mirror about 1-foot in diameter.

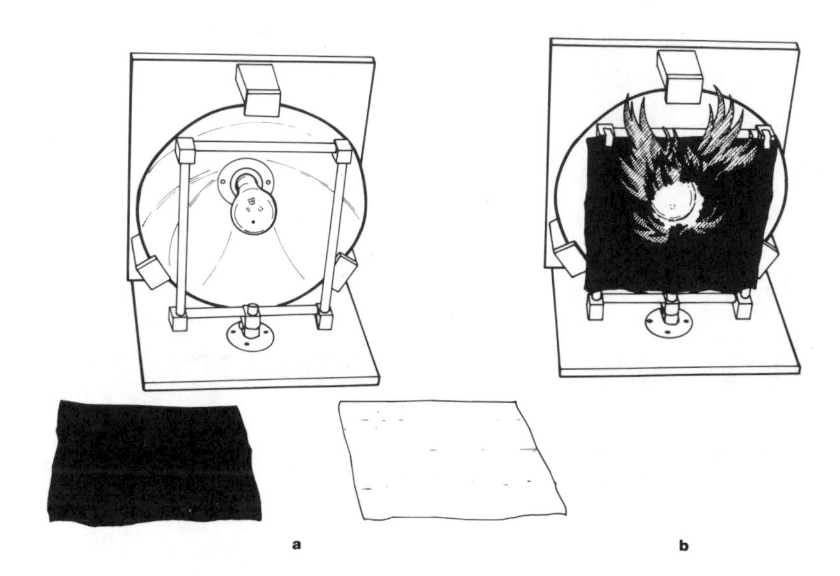

Figure 2.1: Modest apparatus used by the author to demonstrate that electromagnetic radiation possesses energy. (a) Tissue paper, concave mirror, and flashbulb; (b) blackened tissue paper burning (shaded area), concave mirror, and expended flashbulb.

The F/number of the mirror is quite small and consequently the flashbulb is positioned very near the surface of the mirror. I take a *single* sheet of tissue paper and tape it to the frame shown in Figure 2.1a. When the flashbulb is energized, the students see a flash of bright light and nothing happens to the sheet of tissue paper.

The flashbulb is then replaced, and the single sheet of white tissue paper is replaced with a single sheet of tissue paper that has been blackened (and allowed to dry thoroughly) using flat black spray paint. The single sheet of blackened tissue paper is now attached to the frame, and the flashbulb is energized. This time the piece of tissue paper bursts into flames [see Figure 2.1b]!

As simple as this demonstration is, it never fails to evoke great excitement on the part of the students. Who would have thought that there was this much energy associated with the radiation from a flashbulb? The students also see very vividly that blackened surfaces do indeed absorb more visible radiation than white surfaces.

2.2 Searchlight Mirror and Solar Radiation

Our department has a 5-foot diameter mirror that was originally part of an anti-aircraft installation from World War II. These are very useful mirrors, and I would encourage teachers to try to obtain one, perhaps through your state outlet for government surplus properties.

These are all-metal mirrors, i.e., they are not catadioptric systems. Since there is no glass associated with these mirrors, they are quite useful to students for infrared studies of planets, lightning, meteors, and so on. The real image formed of the full moon by this mirror (diameter equals 60 inches, focal length equals 25 inches, F/0.41) is spectacular. They are also useful as "big ears" for work in sound.

I have found such a mirror to be very useful as another way of demonstrating that there is energy associated with electromagnetic radiation. There is something fascinating about a big mirror to most students.

Our physics building is a six-story tower with the sixth floor devoted to most of our astrophysical work. It also contains a porch on which students can set up apparatus. I keep this 5-foot diameter mirror stored on the sixth floor, and with about three students helping me, we roll it out to the porch and *carefully* point it toward the Sun.

Next, I locate in the focal plane of the mirror a piece of wood (2 inches $\times$ 4 inches cross-section) about 5 feet long. The end of

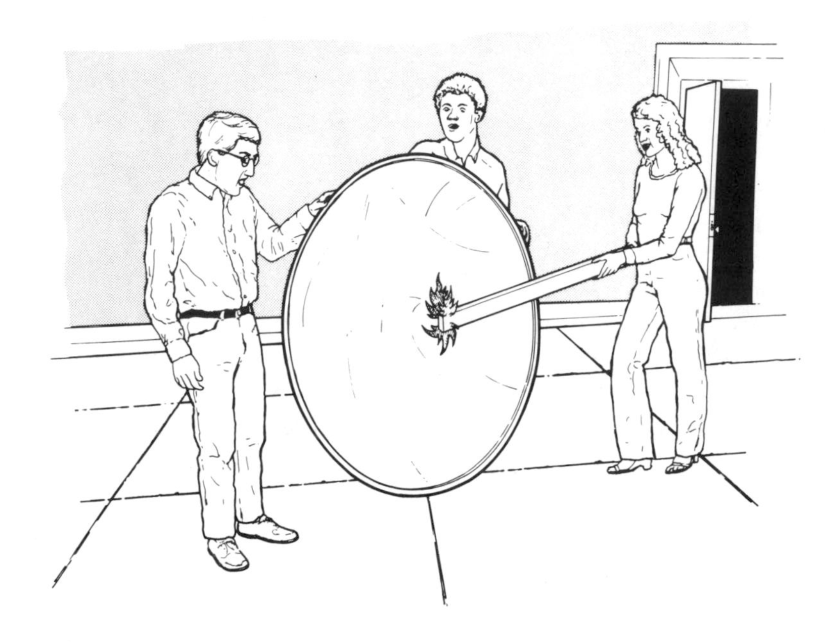

Figure 2.2: Five-foot diameter searchlight mirror and piece of wood used by the author in a beam of solar radiation to demonstrate that electromagnetic radiation possesses energy.

the piece of wood that is placed in the focal plane has previously been blackened using flat black spray paint. Within a matter of seconds the end of the piece of wood in the focal plane will burst into flames (see Figure 2.2)!

Students are told in their texts that the solar constant is about 1400 watts/m^2. This, by definition, is the power density from the Sun at the top of the Earth's atmosphere and is obviously less on the surface of the Earth. Many students have very little feeling for what the solar constant means from an energy point of view. After seeing the wood burst into flames, they have a much better feeling for the numbers involved.

Chapter 3

Planck's Radiation Law

In this chapter we will discuss some relationships that follow from Planck's Radiation Law.

3.1 Stefan-Boltzmann Law and Constant

In Chapter 1 we discussed radiation from objects by examining the Stefan-Boltzmann Law. Historically, this law was discovered empirically before the theory behind it was known. Josef Stefan (1835-1893, Austrian physicist) in 1879, from experimental data, suggested that the rate of emission of radiant energy of all wavelengths from a heated body varies as the fourth power of the absolute temperature. In 1884, Ludwig Boltzmann (1844-1906, Austrian physicist) deduced this relation theoretically for a blackbody. Boltzmann based his deduction upon considerations of the second law of thermodynamics with the aid of the Carnot cycle for an engine. He assumed the operating material to be the radiant energy exerting pressure upon the cylinder walls and the piston. This relationship is known usually as the *Stefan-Boltzmann Law.* However, it is sometimes referred to as the *Boltzmann Law.*

The theory behind the Stefan-Boltzmann Law is Planck's Radiation Law. Max Karl Ernst Ludwig Planck (German physicist) was born in 1858 and died in 1947. He studied under Gustav Kirchhoff and Hermann von Helmholtz. He was also influenced by the work of Rudolf Clausius. At this time in the development of physics, i.e., about 1900, it was easily understood why Planck would become interested in the problem of the nature of the radiation from an object. Planck was in Berlin at this time and so were O. Lummer and E. Pringsheim, who were involved in making measurements of the radiation coming from a small hole in a hot metal cavity, i.e., from a blackbody. These measurements by Lummer and Pringsheim were being made at the Physikalisch Technische Reichsanstalt. Planck realized that he had a profound problem under consideration because Lummer and Pringsheim were finding that the radiation distribution, i.e., the output from the radiation cavity, was independent of the material of which the cavity walls were made and depended only on the absolute temperature of the walls. Planck solved this problem by empirical arguments guided by his mastery of thermodynamics. He considered the radiation in the cavity to be a gas of photons in thermal equilibrium with the cavity walls. In order to solve this problem, Planck made the radical assumption that in the world of the very small, i.e., in the micro world, the energy of oscillators (in particular, the energy of the oscillators making up the walls of the cavity) cannot assume any value but only certain values. This assumption of Planck's suddenly introduces the idea of "graininess" into natural philosophy, in this case a graininess to energy. He postulated that the energy of an oscillator in the cavity walls can have only those energies given by $h \times$ frequency, where h was to be chosen so that his radiation distribution fit the observations of Lummer and Pringsheim. The quantity h is called *Planck's constant.*

Although Planck had been able to arrive at the equation for the spectral radiant emittance of an object, he was deeply concerned with his method of solution. In a letter to R. W. Wood

(Professor of Physics, The Johns Hopkins University) he called his introduction of the constant h "an act of desperation." Before Planck could really understand the theoretical basis for his introduction of *quantization* he had to wait a few years until Louis de Broglie had introduced the concept of dualism (i.e., wave and particle aspects) of matter and Erwin Schroedinger had introduced his quantum mechanical wave equation.

On December 14, 1900, Planck announced the derivation of his equation based on the revolutionary idea that the energy emitted by an oscillator is quantized. He published his derivation in *Annanlen der Physik* **4**, 553 (1901) in a paper entitled "Ueber das Gesetz der Energieverteilung im Normalspectrum" (On the Law for the Distribution of Energy in the Standard Spectrum).

In 1918, Planck won the Nobel Prize in Physics. His Nobel citation read as follows:

> In recognition of the services rendered to the advancement of Physics by his discovery of energy quanta.

For those readers who would like to pursue in greater detail the development of Planck's equation, there are many texts to which you can refer. Four texts are listed below for your convenience, namely,

1. *Introduction to Modern Physics*, by Richtmeyer and Kennard (McGraw-Hill, 1942).
2. *Thermal Physics*, by Morse (Benjamin, 1962).
3. *Quantum Mechanics of Atoms, Molecules, Solids, Nuclei, and Particles*, by Eisberg and Resnick (Wiley, 1974).
4. *Introduction to the Structure of Matter*, by Brehm and Mullin (Wiley, 1989).

Planck's Radiation Law is given by

$$W(\lambda, T) = \frac{c_1}{\lambda^5} \frac{1}{\left[e^{\frac{c_2}{\lambda T}} - 1\right]} \frac{\text{watts}}{\text{cm}^2(\Delta\lambda)}, \tag{3.1}$$

where c_1 and c_2 are referred to as the first and second radiation constants, respectively, λ is the wavelength, and T is the temperature of the source in Kelvin. In this equation, W is referred to as the *spectral radiant emittance.*

Figure 3.1 has been included to help the reader with this law and from this figure the reader will see that the first and second radiation constants have been expressed in terms of the velocity of light, c, the Planck constant, h, and the Boltzmann constant, k. Figure 3.1 requires careful observation and study because it contains much information.

As indicated in this figure, if one integrates Planck's equation from "0" to "∞", i.e., integrates over all wavelengths, one obtains the Stefan-Boltzmann Law. Note that the Stefan-Boltzmann constant can be expressed in terms of c, k, and h.

3.2 Wien Displacement Law

For a fixed T, W becomes a function only of λ, as shown in Figure 3.1. From this figure, one notes that at the peak of the curve $dW/d\lambda$ equals "zero." Calling the wavelength corresponding to the peak of the curve λ_{maximum}, carrying out the differentiation referred to previously, and equating it to zero, one obtains the following equation:

$$\lambda_{\text{maximum}} T = 2890 \ \mu\text{m degrees.} \tag{3.2}$$

This is known as the *Wien Displacement Law* and was discovered empirically before it was understood theoretically.

The reader is urged to study carefully the Wien Displacement Law and to try to appreciate the fact that it is a powerful tool in remote sampling. Imagine, if you will, an infrared spectrometer located in the focal plane of a reflecting telescope and pointed toward a distant object, say a star. The spectrometer, if it has been wavelength and radiometrically calibrated (see Chapter 8 for how this is done), will yield a plot similar to the curve shown

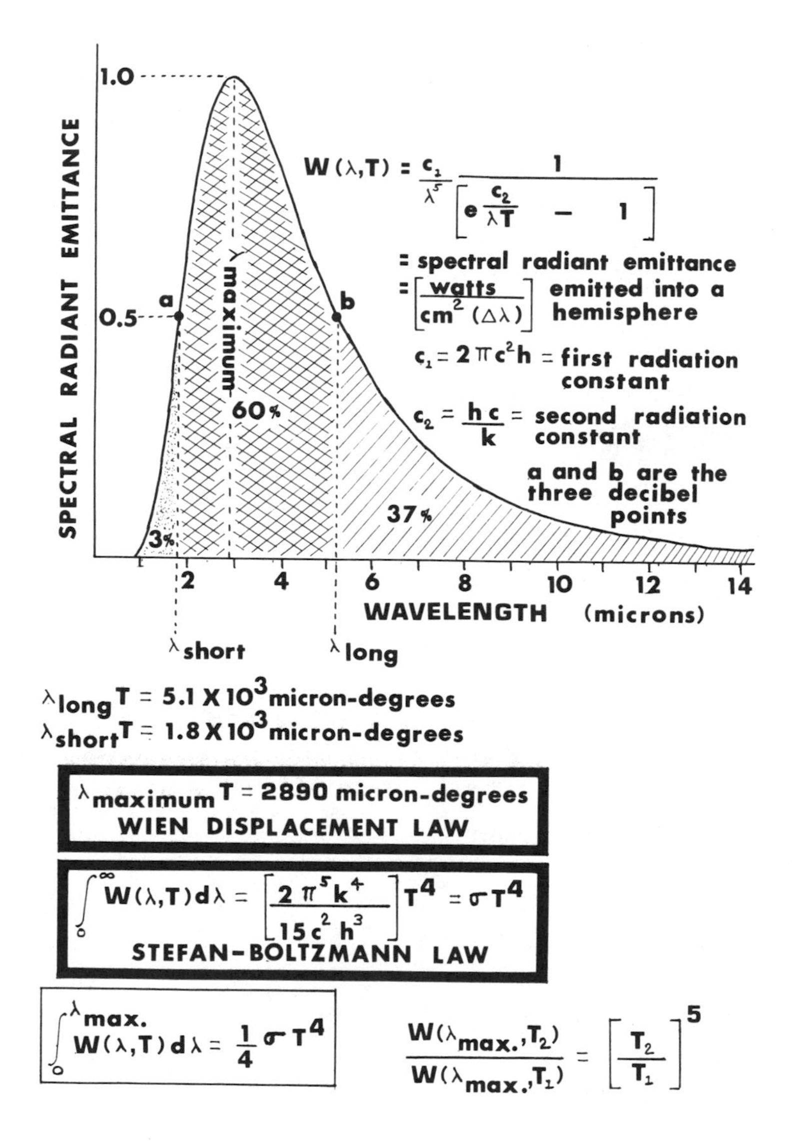

Figure 3.1: Planck's Radiation Law.

in Figure 3.1. By measuring λ_{maximum} and invoking the Wien Displacement Law one can determine the temperature of the star!

Many relationships can be developed from the Planck Radiation Law. For example, the peak radiation from two objects at different temperatures, say one body at 1000 K and another body at 500 K, will vary as the ratio of the absolute temperatures raised to the fifth power, i.e., $(1000/500)^5 = 32$. If one normalizes the peak of the radiation curve to 1 and comes down on the ordinate scale to the point corresponding to 0.5, and at that point draws a horizontal line, a very interesting result can be obtained.

3.3 Relationship Involving the "Three Decibel Points" of the Planck Radiation Law

At the points where the horizontal line crosses the Planck radiation curve, i.e., at points *a* and *b*, drop perpendiculars to the wavelength axis. Call the points where these perpendiculars cross the wavelength axis λ_{short} and λ_{long}, as indicated in Figure 3.1. One can easily show, although we shall not do it, that between λ_{short} and λ_{long} there is 60% of the total area under the curve. Also one can show that from λ_{long} toward longer wavelengths there is 37% of the area under the curve and that from λ_{short} to shorter wavelengths there is 3% of the area of the curve. One can also show that from zero to λ_{maximum} there is 25% of the total area under the curve.

One can work out other relationships similar to the Wien Displacement Law that indicate the following:

$$\lambda_{\text{short}} T = 5.1 \times 10^3 \mu\text{m degrees}; \quad (3.3)$$

$$\lambda_{\text{long}} T = 1.8 \times 10^3 \mu\text{m degrees}. \quad (3.4)$$

The various relationships indicated in Figure 3.1 can be very useful for computational purposes. The reader is urged to try to develop a feeling for the differential nature of Planck's Radiation Law [Equation (3.1)]. $W(\lambda, T)$ can have a finite value only when one considers a finite wavelength interval.

From a historical standpoint, the two earliest attempts at explaining curves such as shown in Figure 3.1 were the empirical radiation laws of Wien and Rayleigh-Jeans. These laws are, of course, special cases of the more general law developed by Max Planck.

3.4 Empirical Radiation Laws

3.4.1 Wien Radiation Law

We shall now show that Equation (3.1) takes on simpler forms for small and large values of the product, λT. For λT small, $e^{\frac{c_2}{\lambda T}} >> 1$ and Equation (3.1) becomes

$$W(\lambda, T) = \frac{c_1}{\lambda^5}\left[\frac{1}{e^{\frac{c_2}{\lambda T}}}\right]. \tag{3.5}$$

Equation (3.5) is known as the *Wien Radiation Law.*

Wilhelm Wien (1864-1928, German physicist) in 1891 derived his displacement law using the same approach as that used by Stefan and Boltzmann. This approach, however, was not able to lead to a solution of the variation of radiant intensity with wavelength and temperature of the blackbody. By the use of additional principles, Wien was able to obtain Equation (3.5).

The Wien Radiation Law applies if λT is less than 0.3 cm degrees. This means that it applies well in the visible spectrum for source temperatures up to about 4000 K. Assuming an upper limit to the visible spectrum of 0.7 μm, then (0.7 μm)(4000 K) = 0.28 cm degrees. It should be pointed out that the Wien Radiation Law is quite useful in optical pyrometry. Wien won the Nobel prize in physics in 1911 for his work on the radiation laws.

3.4.2 Rayleigh-Jeans Radiation Law

Let $x = c_2/\lambda T$ and write Equation (3.1) as follows:

$$W(\lambda, T) = \frac{c_1}{\lambda^5}\Big[\frac{1}{(e^x - 1)}\Big]. \tag{3.6}$$

Recall the series expansion for e^x, namely,

$$e^x = 1 + x + \frac{x^2}{2!} + \frac{x^3}{3!} + \frac{x^4}{4!} + \ldots. \tag{3.7}$$

Equation (3.6) can then be written as

$$W(\lambda, T) = \frac{c_1}{\lambda^5}\frac{1}{\Big[(1 + x + \frac{x^2}{2!} + \frac{x^3}{3!} + \frac{x^4}{4!} + \ldots) - 1\Big]} \tag{3.8}$$

Assume x to be small (less than unity) and neglect higher terms in x^2, x^3, x^4, etc. In other words, assume λT is large. Under this assumption, Equation (3.8) becomes

$$\begin{aligned} W(\lambda, T) &= \frac{c_1}{\lambda^5}\frac{1}{[1 + x - 1]} \\ &= \frac{c_1}{\lambda^5}\Big[\frac{1}{x}\Big], \qquad (3.9) \\ W(\lambda, T) &= \frac{c_1}{\lambda^5}\Big[\frac{\lambda T}{c_2}\Big], \\ W(\lambda, T) &= \frac{c_1}{c_2}\Big[\frac{T}{\lambda^4}\Big]. \qquad (3.10) \end{aligned}$$

Equation (3.10) is known as the *Rayleigh-Jeans Radiation Law.* Lord (John William Strutt) Rayleigh (1842-1919, British physicist) and Sir James Hopwood Jeans (1877-1946, British physicist and mathematician), also using thermodynamic reasoning, were able to derive Equation (3.10). This radiation law breaks down at short wavelengths. It suffers from what is called "the ultraviolet catastrophe." The Rayleigh-Jeans Radiation Law is valid if λT is much greater than 77 cm degrees. For a 1000 K source, this condition could be satisfied for wavelengths greater than 770 μ m. This radiation law applies quite well in the far infrared and microwave regions of the electromagnetic spectrum. Lord Rayleigh received the Nobel prize in physics in 1904 for the discovery of argon.

Chapter 4

Kirchhoff's Radiation Law

4.1 A Simple Demonstration

It was Gustav Robert Kirchhoff (1824-1887) who pointed out in 1860 that the radiation issuing from a suitably small hole in a uniformly heated enclosure is approximately blackbody (see Chapter 5) radiation and who discovered the universal character of blackbody radiation. He showed that the radiation from a blackbody is a function of the temperature alone and is independent of the material of which the enclosure is made.

Kirchhoff's Radiation Law is often vulgarized to "good absorbers are good emitters of radiation." However, it must never be forgotten that the identity of the spectral radiant absorptance and the spectral radiant emittance only holds when both refer to the same wavelength and temperature. This law follows from the law of conservation of energy and from thermal equilibrium considerations. It should be pointed out that it follows from Kirchhoff's Radiation Law that the polar diagram for the transmission and reception of radio signals must be the same. A straightforward way to demonstrate this is to use a Leslie cube filled with hot water, a radiation thermopile, and a galvanometer. This apparatus is shown in Figure 1.4.

A Leslie cube is a thin-walled metal can (in the shape of a

cube) with an opening in the top for pouring the hot water. Of the four vertical sides, one is shiny and the other three are painted different colors, one of which is black. Since the temperature of all faces of the cube is the same, any difference in the amount of radiation (i.e., the radiance) is due to a difference in the radiant emissivity of the face of the cube.

When the Leslie cube is positioned in front of the radiation thermopile (I use a lab jack, on top of which a small piece of wood is placed for supporting the cube), one records a considerably larger galvanometer deflection from the blackened side of the cube than from the shiny side of the cube. Obviously, the blackened side of the cube is a better absorber of visible radiation than is the shiny side. It is also a better absorber in the infrared.

4.2 An Elegant Demonstration

Unless the temperature of an object is quite high, the majority of its radiation will lie in the infrared spectral region. By having an apparatus that permits infrared spectroscopic measurements, one will find that whole new vistas of opportunities will have been opened.

Figure 4.1 is a sketch of the experimental setup used by the author. It should be noted that this arrangement is mounted on casters, so the apparatus can be easily rolled to wherever it might be needed. There is even a water pump as part of this system so that water for use with the globar source is readily available at all locations.

The spectrometer sketched in Figure 4.1 is a Perkin-Elmer model 12 C infrared spectrometer with its associated electronics. The strip chart recorder is a Leeds & Northrup type G. I have found the Perkin-Elmer model 12 C to be a superb teaching instrument. These spectrometers have been commercially available for many years, certainly since the 1940s.

With the availability of an apparatus such as that sketched

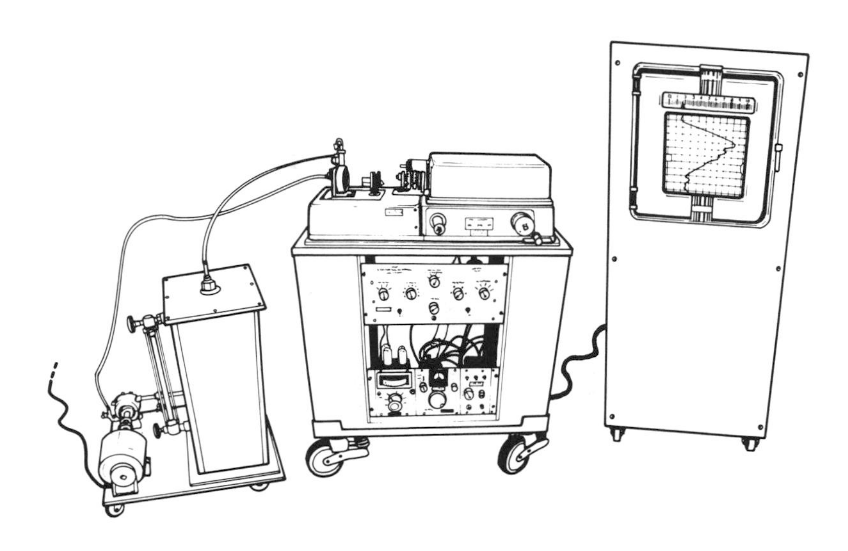

Figure 4.1: Another setup to demonstrate Kirchhoff's Law dealing with radiation, using an infrared spectrometer and a strip chart recorder.

in Figure 4.1, one can carry out what I call an elegant demonstration of Kirchhoff's Radiation Law. Figure 4.2 is a schematic drawing of the demonstration. As shown in Figure 4.2(a), one first observes the absorption spectrum of carbon dioxide. One uses a globar source (a rod of bonded silicon carbide through which current is passed in order to bring its temperature to about 1227°C or 1500 K) and an absorption cell (about 10 cm in length) equipped with sodium chloride windows for this observation. One obtains a trace on the strip chart recorder, as indicated in Figure 4.2(a).

To carry out the next part of the demonstration, one needs to make a helix of copper tubing (about 3/8 inch outside diameter and 1/4 inch inside diameter) that is several feet in length. One

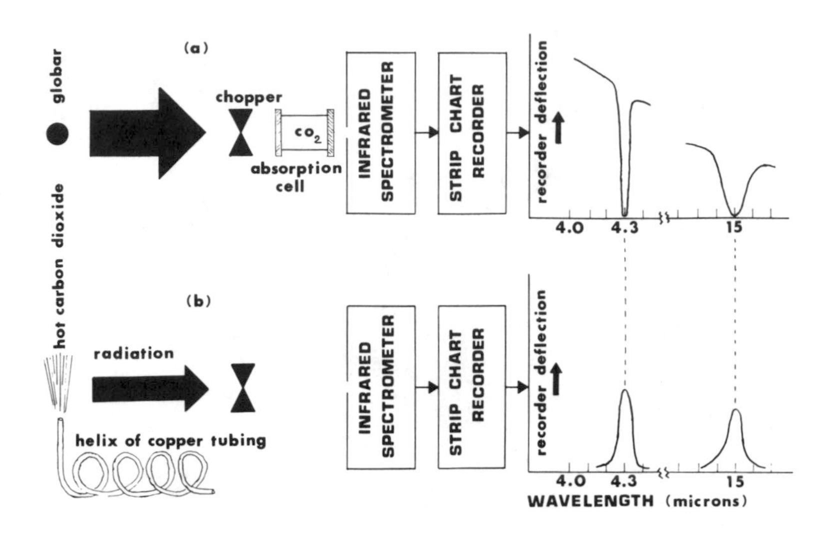

Figure 4.2: Schematic drawing of the demonstration using the infrared spectrometer. (a) The absorption spectrum of carbon dioxide. (b) The deflection produced on the strip chart recorder.

then arranges to pass carbon dioxide gas through the helix while it is being heated by several Bunsen burners. The escaping hot carbon dioxide gas now becomes the source of radiation instead of the globar.

In the spectrometer shown in Figure 4.1, I have modified the base by drilling and tapping two 1/4–20 holes in order to be able to locate the radiation chopper in front of the entrance slit, in addition to its usual location in front of the globar. This modification, plus the rotation of a plane mirror in the source housing of this spectrometer, makes it a simple matter to image the radiation from the hot carbon dioxide gas on the entrance slit of the spectrometer.

When one now scans the spectrum, the deflection on the strip chart recorder will be as indicated in Figure 4.2(b). This is an

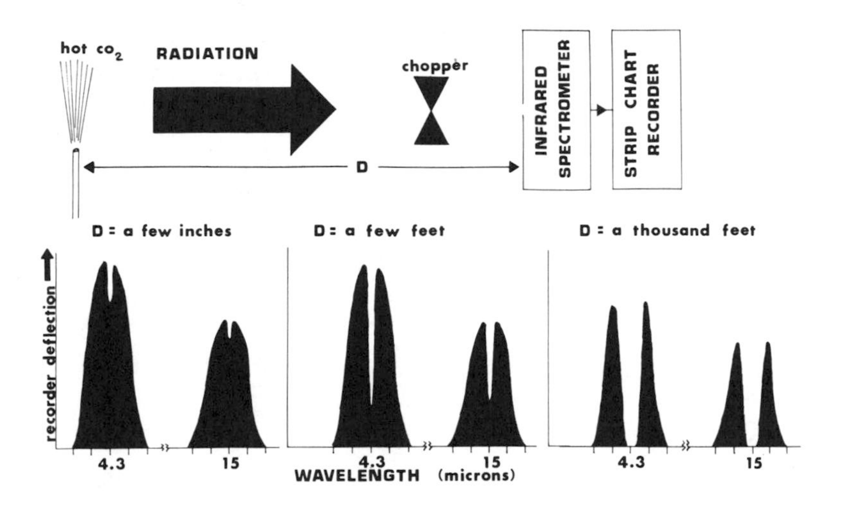

Figure 4.3: Demonstration of self-absorption, showing the result of increasing amounts of absorption by carbon dioxide in the radiation path.

extremely pedagogical point. Thus, we observe, in keeping with Kirchhoff's Radiation Law, that carbon dioxide emits in those spectral regions where it absorbs, namely, at 4.3 μm and 15 μm.

4.3 Self-Absorption

With a little additional effort, one can extend this elegant demonstration of Kirchhoff's Radiation Law to demonstrate the phenomenon of self-absorption. Figure 4.3 is a sketch of this demonstration. To carry out this part of the demonstration, one will need a first-surface aluminized spherical mirror in order to form a real image of the radiation from the hot carbon dioxide gas at the same distance that the globar is from the entrance slit of the spectrometer.

By changing the distance between the hot carbon dioxide

and the spectrometer, one causes the radiation to travel over greater distances and hence through more carbon dioxide gas. The result of increasing amounts of absorption by the carbon dioxide is shown in Figure 4.3.

The case corresponding to a distance of about 1000 feet is quite interesting in that what began as an "emission band" now appears to be two separate emission bands. The phenomenon of self-absorption is often encountered in carrying out spectroscopic interpretations.

Chapter 5

Radiant Emissivity

5.1 Metals

It is important to note that the radiant emissivity of both metals and gases can be changed. Let us begin with a discussion of metals, using Figures 5.1 and 5.2 to assist in our discussion. Much material has been included in these figures, and the reader is encouraged to study them carefully. Many arrows are shown, and their size should be interpreted as being proportional to the radiant emissivity of the object concerned.

Let us begin by considering Figures 5.1(a) and (b). By simply bending a piece of metal, one can enhance its radiant emissivity. Such a bend, as indicated in Figure 5.1(b), is referred to as a *Mendenhall wedge.* By forming a wedge one is causing incident radiation to undergo more reflections, and hence more absorption, and hence approaching more and more closely what is called a *blackbody* or *hohlraum.*

Hohlraum is the German word for blackbody. By definition, a blackbody is an object that absorbs all incident electromagnetic radiation, i.e., it reflects none of the incident electromagnetic radiation. As the object becomes a better absorber, it also becomes a better emitter. This, of course, is Kirchhoff's Radiation Law.

Other approaches to this bending of the metal, and hence

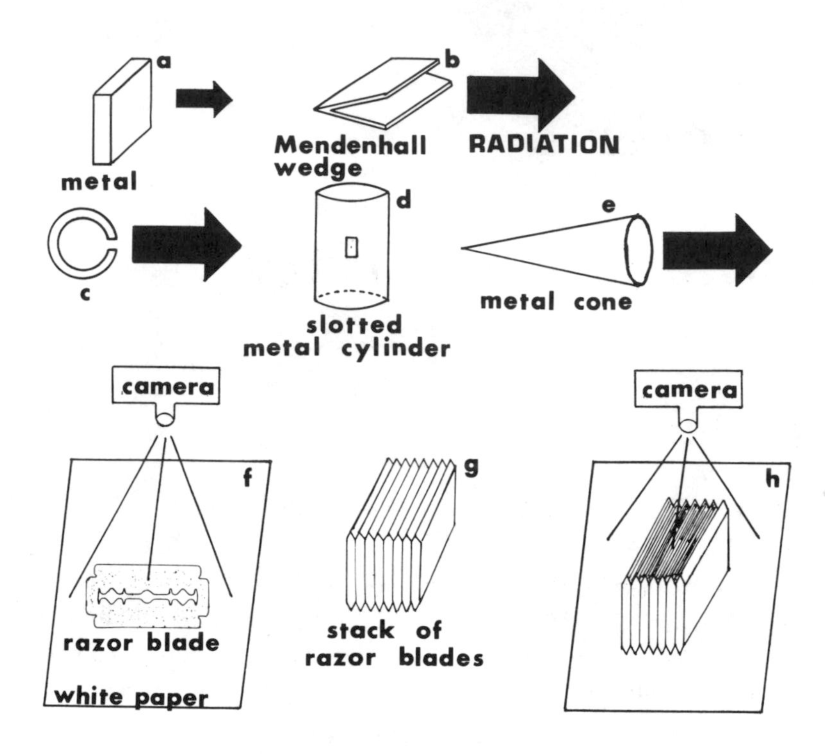

Figure 5.1: Radiant emissivity of metals.

trapping of the radiation, are shown in Figures 5.1(c), (d), and (e). Although we are discussing the shaping of metals, the reader should pay close attention to the appearance of the apex in a paper drinking cone the next time you use one. What is the appearance of the apex?

The shape indicated in Figure 5.1(e), namely a cone, is the shape frequently used in the construction of commercial reference radiation sources, i.e., so-called blackbodies. In this case, the cones are made of ceramic material rather than metal. Some of these commercial sources can be operated at temperatures as high as 1200°C. It should be appreciated that the radiant emissivity of an object is a function of the temperature and of the wavelength.

I always suggest to my students that they carry out the experiment indicated in Figures 5.1(f), (g), and (h). First, one takes a double-edged razor blade and places it on a piece of white paper. When the film is developed and printed, one has a print in which there is very little contrast, i.e., both the white paper and the shiny razor blade are good reflectors of visible light.

One next repeats the experiment by stacking together several razor blades, as indicated in Figure 5.1(g). I suggest to the students that they hold the razor blades together by using a rubber band (though this is not indicated on the drawing). One then takes this stack of razor blades and places them edge-on on the white paper, as indicated in Figure 5.1(h). When the photograph is taken and developed, one has a print in which there is now considerable contrast between the edges of the razor blades and the white paper. The Vs between the razor edges are acting like Mendenhall wedges, providing considerable absorption of visible light.

5.2 Gases

Let us now discuss the radiant emissivity of gases, as indicated in Figure 5.2. First, if one were to measure the radiation from a gas (assume carbon dioxide), it would be observed that there was not radiation throughout most of the spectrum, as would be the case with metals, but only those spectral regions where the gas absorbs. This is, of course, simply another manifestation of Kirchhoff's Radiation Law.

One would find that the radiant emissivity could be increased by increasing the density of the gas, by making the column of gas longer, or by both of these methods. If the amount of gas in

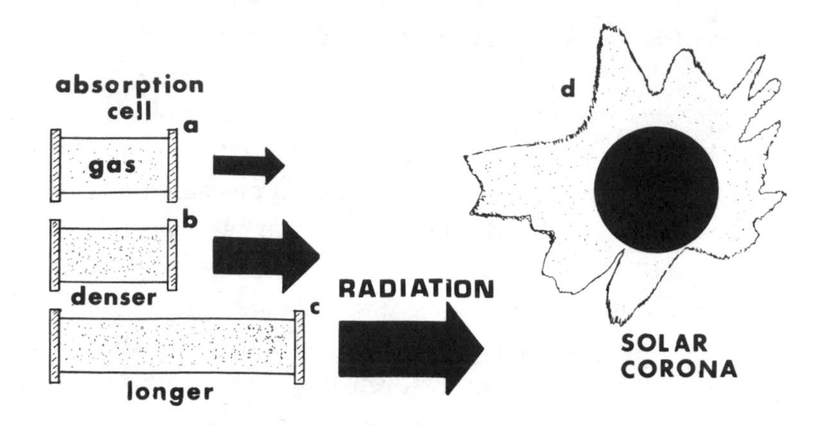

Figure 5.2: Radiant emissivity of gases.

the path is such that the radiation passing through the column is completely absorbed at one or more wavelengths, then the column of gas at those wavelengths has a radiant emissivity of *unity* and is said to be a *blackbody* at those wavelengths.

The reader's attention is invited to the sketch shown in Figure 5.2(d). The solar corona is characterized by having an extremely high temperature but a very low radiant emissivity. It should be pointed out that a radiometer, i.e., a spectroscopic instrument that measures radiation over a fairly broad band of wavelengths, is not able to distinguish an object of high temperature and low radiant emissivity from an object of low temperature and high radiant emissivity.

Part II

Some Applications of Radiation Exchange

Chapter 6

Ice and Space

6.1 Infrared Spectrum of a Block of Ice

The subject of radiation exchange holds great promise for the teacher and student alike in being extremely open-ended. It is laden with astrophysical and space applications, both of which are of interest to most students. In Part II we begin a discussion of some applications of radiation exchange. These demonstrations and experiments range all the way from the infrared spectrum of a block of ice to the measurement of the effective radiating temperature of the ozonosphere.

Unless one is familiar with the Stefan-Boltzmann Law, this particular application might seem difficult to perform. In Chapter 1, we observed the exchange of radiation between a thermopile, a hot soldering iron, and a beaker of ice. In that demonstration, the soldering iron produced a galvanometer deflection that was in the opposite direction to that produced by the ice. You know that the galvanometer deflection associated with the beaker of ice was in the direction it was because the ice was colder than the thermopile. You will also recall that in that demonstration the thermopile responded to the net amount of radiation, either in the case of the beaker of ice or in the case of the soldering iron.

We now want to repeat that demonstration, except now we want to do it more elegantly. We shall use the apparatus shown in Figure 4.1, namely, the Perkin-Elmer model 12 C infrared spectrometer. Instead of observing a very broad band of wavelengths, as in the case of the radiation thermopile, we are now going to observe the radiation from an object in very narrow bands of wavelengths and cause these narrow bands of wavelengths, seriatim, to fall upon our detector, which in this case is a radiation thermocouple. This, of course, is what a spectrometer is designed to do. This particular spectrometer is equipped with a sodium chloride prism.

Figure 6.1 is a schematic drawing of this demonstration. It will be noted in Figure 6.1(a) that we record first the spectrum of an object that is hotter than the thermocouple, in this case a rod of globar that is incandescent. Next, we locate a block of ice at the same distance from the entrance slit of the spectrometer as the globar was [see Figure 6.1(b)]. We will have formed a cone-shaped opening in the block of ice. In carrying out this demonstration, one records first the spectrum of globar. One then rolls back the strip chart recorder paper and starts at the beginning of the recorder paper with the spectrum of ice. Figure 6.1(c) is a sketch of what one obtains from the strip chart recorder after certain computations have been made.

I think we can get through this part without becoming mired down in technical detail. In order to do this, one needs to have carried out what is called a *wavelength calibration of the spectrometer*. This is a straightforward operation. It is the next step in the operation where the technical difficulty is likely to arise.

To be able to measure *spectral radiant emittance* as indicated in Figure 6.1(c), one needs to know how many watts of radiation striking the thermocouple produce how many units of deflection on the recorder paper. It is necessary to use a reference radiation source (or blackbody) for this phase of the radiometric calibration.

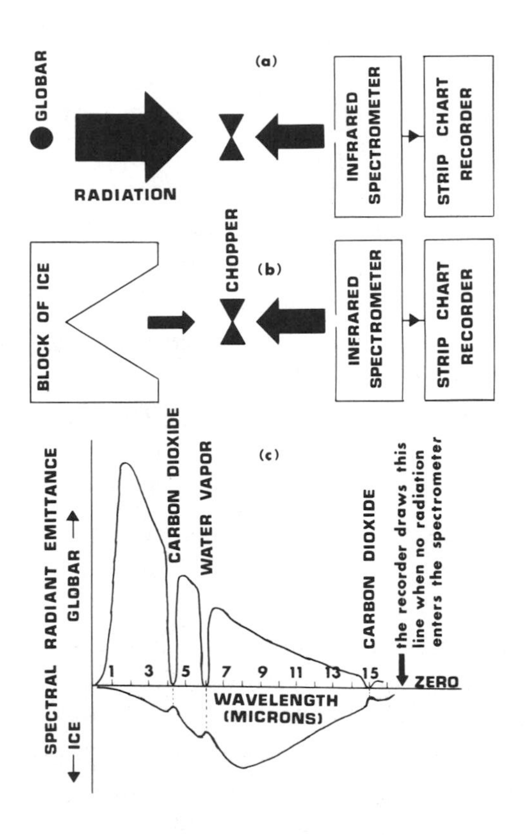

Figure 6.1: Infrared spectrum of ice.

Watt/μm/cm^2 are the units for spectral radiant emittance radiated into a hemisphere. Knowing this calibration factor, one then measures the deflection on the chart paper throughout the spectrum and, by using the calibration constant, is able to draw a curve such as the one shown in Figure 6.1(c).

Try not to be confused by this technical phase. Instead, observe closely the information shown in this figure. One sees that when the object is hotter than the thermocouple, the deflection produced is in a certain direction and that when the object is colder than the thermocouple, the deflection is in the opposite direction as shown in the sketch.

We note in the globar trace that carbon dioxide and water vapor in the atmospheric path between the globar and the thermocouple produce absorption. If one notices carefully the downscale deflection produced by the ice, there are certain places where this downscale deflection suddenly goes upscale. In other words, the deflection goes upscale at those places where there is more radiation. There is more radiation exactly at those regions of the spectrum where there is more absorption. Again, this is another example of Kirchhoff's Radiation Law.

The author cannot help but comment on the "downscale" or "negative" deflection shown in Figure 6.1(c). I had taken a group of undergraduates to Gulkana, Alaska to study the total solar eclipse of 1963. Several different experiments were performed, and one of them involved the use of a thermistor bolometer detector in a large radiometer.

After the extensive expedition was finally set up and we were pointing the various optical systems at different objects to make sure everything was in order, I suggested to the two students who were tracking the large radiometer that they point it at one of the mountain peaks that could be seen from our location. Suddenly a negative signal appeared on the strip chart recorder and there was considerable confusion. They had failed to realize they were looking at snow on the top of the mountain and that it was colder than their bolometer.

6.2 Crookes' Radiometer in a Block of Ice

Figure 1.4 shows a sketch of a Crookes' radiometer, an apparatus that is traditionally included in radiometric demonstrations. It consists of a small paddle wheel with four very thin vanes that have been blackened on one side and highly polished on the other side.

When radiation, say from a tungsten lamp, falls upon the vanes, the paddle wheel revolves. The simplest explanation of this phenomenon is that the *net* radiation exchange warms the blackened sides slightly more than the shiny sides. As a result of this unequal warming, air molecules rebound more vigorously from the blackened sides than from the shiny sides and the paddle wheel revolves because of this differential reaction, i.e., the blackened sides move away from the tungsten lamp. It is important to have the pressure in the bulb correct.

Although the author has never actually performed the following demonstration, he always discusses the problem with students, namely, what would happen to the direction of rotation of the paddle wheel if the Crookes' radiometer were surrounded by an object that is colder than the vanes of the radiometer. Since the blackened vanes have a higher radiant emissivity than the shiny vanes, one predicts that the *net* radiation exchange is such that the blackened vanes will cool more than the shiny vanes. As a result of this unequal cooling, air molecules now rebound more vigorously from the shiny vanes than from the blackened vanes, and the paddle wheel revolves in the opposite direction as a result of this differential reaction.

Instead of using cakes of ice that have been scooped out to surround the radiometer, one could also use dry ice. Another possibility would be to bring a small refrigerator to the lecture room and place the radiometer in the freezing compartment. Yet another possibility would be to build an appropriately shaped double-walled box, permitting one to load the space between the walls with crushed ice.

For those readers who are interested in some additional material on the subject, let me refer you to the article by Frank S. Crawford entitled "Running Crookes' Radiometer Backwards" that appears on page 490 of the June 1986 issue of the *American Journal of Physics*.

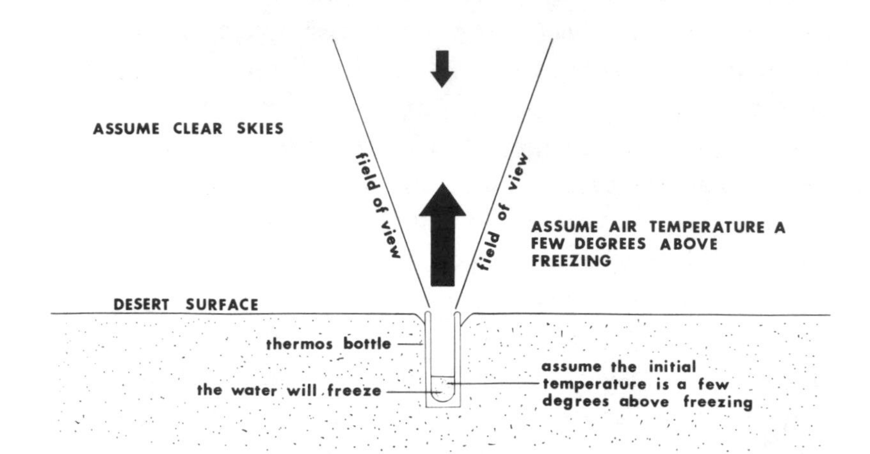

Figure 6.2: Schematic drawing of radiation exchange taking place at nighttime on the desert.

6.3 Nighttime on the Desert

From what we have already discussed in this book, one now has sufficient background on the phenomenon of radiation exchange to appreciate that if one were caught on the desert at night without adequate clothing, the situation might very well become critical.

Let us examine the situation sketched in Figure 6.2. The relative humidity on the desert is likely to be quite low, indicating a small amount of water vapor in the atmosphere and hence less absorption of the infrared radiation leaving the Earth. We assume clear skies and that the air temperature is a few degrees above freezing. We want to place a thermos bottle (open on the top) with some water in it on the desert at night. Since the water temperature is considerably higher than the temperature of space (note: we are assuming clear skies), one observes, as a result of radiation exchange, that the water in the thermos bottle will freeze!

Many man-made satellites in orbit around the Earth are equipped with infrared radiation detectors because many objects of interest have optical signatures in the infrared. It is necessary to cool many of these detectors. By allowing infrared detectors to exchange radiation with space, H. W. Yates and his co-workers at the National Environmental Satellite Center have been able to cool the detectors to about 100 K.

Chapter 7

Planetary Studies

7.1 Measuring the Effective Radiating Temperature of the Ozonosphere

The experiment described here represents an application of radiation exchange. The center of the ozonosphere lies about 15 miles above the surface of the Earth. In this discussion, we are not going into the photochemical reactions taking place in the ozonosphere, but instead shall focus on measuring its effective radiating temperature.

This experiment requires infrared spectroscopic equipment plus ancillary optical equipment consisting of a first-surface paraboloidal mirror, one or more plane first-surface mirrors, and a coelostat or heliostat for tracking the Sun. The mechanical part of the tracking system could be built in a machine shop.

Since the ozone layer does not have a constant temperature, i.e., it is not an isothermal layer, one must be content with a measurement of what is called the *effective radiating temperature.* The ozone molecule has several absorption bands in the ultraviolet, visible, and infrared regions. This experiment concerns only the infrared absorption. In the infrared there are absorption bands located at 4.7, 9.6, and 14.2 μm. The 9.6 μm band is chosen for this experiment for several reasons, namely,

- the band at 4.7 μm is contaminated by other atmospheric absorptions,
- the band at 14.2 μm is almost completely obscured by carbon dioxide and water vapor absorption,
- the radiant absorptance of the 9.6 μm band is greater than that for any other band of the ozone molecule,
- the temperature of the ozonosphere is such that its peak radiation is near 9.6 μm, and
- the atmosphere is relatively free of absorption by carbon dioxide and water vapor in the 9.6 μm region.

In this experiment, one observes the radiance from a column of ozone. Since the column does not have sufficient optical thickness to be a blackbody, it is necessary to determine its radiant emissivity at 9.6 μm. Many of the details for carrying out this experiment are deliberately omitted. Instead, we shall zero in on the physics involved.

As indicated in Figure 7.1, one uses the coelostat system to track the Sun and scan a spectral region on either side of 9.6 μm [see + signal on recorder deflection in Figure 7.1]. From this spectrum, one determines the radiant absorptance at 9.6 μm as follows:

$$\text{Radiant absorptance} = \text{s}/\text{S} = \text{Radiant emissivity}. \qquad (7.1)$$

From Kirchhoff's Radiation Law, we know the radiant emissivity is equal to the radiant absorptance.

Note carefully the various radiations that are involved in Figure 7.1. Not only is there radiation from the Sun, but there is also radiation from the ozonosphere and from the blackened thermocouple in the spectrometer. These various radiations are

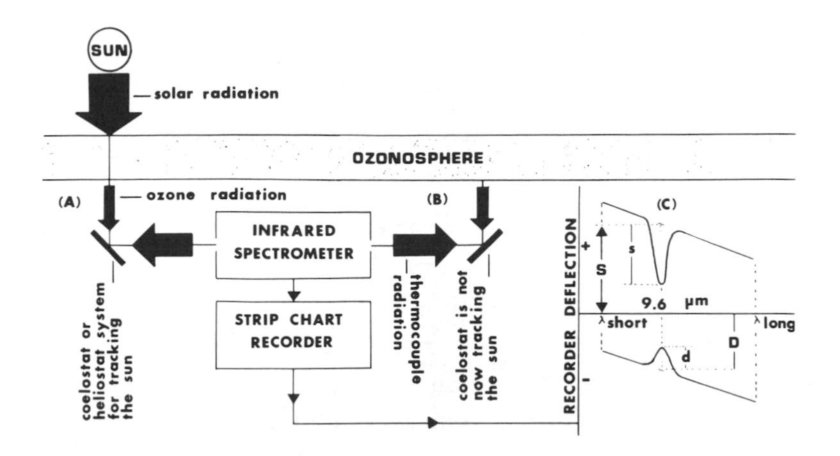

Figure 7.1: Illustration of the key ideas involved in the measurement of the effective radiating temperature of the ozonosphere. In this figure, a = spectrum of the Sun, and b = spectrum of the ozonosphere.

indicated by the blackened arrows. In the case of the Sun, the radiation from the ozonosphere and from the blackened thermocouple can be disregarded.

After recording the 9.6 μm region in absorption, one stops tracking with the coelostat and lets the Sun move out of the spectrometer's field of view. Also, the coelostat primary mirror is turned back to a position that enables the spectrometer to "see" the same region of the ozonosphere used to measure the radiant absorptance. One now records the exchange of radiation between the column of the ozonosphere and the blackened thermocouple, as indicated in Figure 7.1 [see – signal on recorder deflection in Figure 7.1]. It is important to note that the signal at 9.6 μm is less negative than in nearby spectral regions. This, of course, is simply another example of Kirchhoff's Radiation Law. One will need to measure d/D from the spectrum.

In recording the two spectra sketched in Figure 7.1, it is imperative that the width of the entrance slit of the spectrometer be the same, i.e., the spectral resolving power must be the same. In order to do this, it will be necessary to attenuate the radiation coming from the Sun. One does this by inserting a large attenuator in the parallel beam of solar radiation. This attenuator is made from a piece of aluminum about two mm thick and about one meter by one meter in size. Imagine lines spaced about 20 mm apart drawn parallel to the edges of this plate. At the intersections of these lines holes measuring about one mm in diameter are drilled through the aluminum plate.

In determining the effective radiating temperature of the ozonosphere, the assumption is made that the thermocouple radiates as a blackbody in the region of 9.6 μm. Using a thermometer located in the spectrometer, one can determine the temperature of the thermocouple and, hence, can theoretically calculate the radiance from the thermocouple in the 9.6 μm spectral region. Let us call the radiance of the thermocouple R. The following expression becomes the working equation for the determination of the effective radiating temperature T of the ozonosphere, namely,

$$R(S/s)(d/D) = \int_{\lambda}^{\lambda+\Delta\lambda} \frac{c_1}{\lambda^5}\left[\frac{1}{e^{\frac{c_2}{\lambda T}} - 1}\right] \Delta\lambda \, \frac{\text{watts}}{\text{cm}^2\,\text{micron}}. \quad (7.2)$$

This experiment was originally done by Arthur Adel and was published in December 1949 in *Air Force Cambridge Research Laboratories' Geophysical Research Paper No. 2.*

7.2 The Greenhouse Effect: Radiant Transmittance of Glass

The equipment sketched in Figure 1.4 is used to confirm some of the ideas that are involved in a discussion of the theory of the greenhouse effect. With the energized soldering iron placed several feet in front of the radiation thermopile, one obtains full-scale deflection on the galvanometer. A lab jack is then used to support a piece of high quality optical glass (about one inch thick) that is large enough to completely cover the reflecting cone associated with the radiometer. The students observe the galvanometer deflection return to zero, indicating that glass does not transmit infrared. In other words, radiation beyond about 2 μm is absorbed by the glass. This observation is very critical in a discussion of the greenhouse effect.

Viewing Figure 3.1 on the Planck Radiation Law, imagine such a curve drawn for an object with a temperature of 6000°C (approximately the surface temperature of the Sun). The peak of such a curve would lie in the visible region of the spectrum. In fact, almost 60% of the radiation from the Sun lies in the visible.

Since glass is transparent in the visible, the majority of the radiation from the Sun gets through the greenhouse glass and is absorbed by the plants and other contents inside. The temperature of the objects in the greenhouse is about 23°C or roughly 300 K. Invoking the Wien Displacement Law, we calculate that the peak radiation from such objects will take place at about 10 μm. However, since glass is not transparent at 10 μm, the radiation from the objects inside the greenhouse is absorbed by the glass. The temperature of the glass will rise and the glass will radiate more. Part of this radiation is toward the outside of the greenhouse, and part of it is toward the inside. It is this latter radiation that helps to hold up the temperature inside.

7.3 Surface Temperature of Planet Earth

7.3.1 Atmospheric Transmittance: The Concept of Atmospheric Windows

This is a topic of considerable importance for all of us. It is almost too much to accept the fact that we are not getting too hot and we are not getting too cold. How is it that these two tendencies are just balanced? In attempting to come to grips with this problem, the reader will have the opportunity to use many of the ideas we have discussed in this book.

At the outset one will recognize certain similarities between this problem and that of the greenhouse effect previously discussed. In this problem, the glass is replaced by certain minor constituents in the Earth's atmosphere that absorb in certain wavelength regions. One needs to appreciate that the energy that drives the Earth is solar radiation and that approximately 60% of this solar radiation lies in the visible. Furthermore, these minor constituents that absorb do not absorb solar radiation lying in the visible.

We have already seen that the important consideration in matters involving radiation is the *net radiation* involved. As far as the side of the Earth facing the Sun is concerned, there is no doubt about the direction of the net radiation exchange. The Earth is radiating to the Sun, but the amount of radiation coming from the Sun is considerably greater and the net radiation is in toward the Earth. Since the Earth is rotating, different parts are constantly being exposed to solar radiation. The part of the Earth that cannot "see" the Sun will, of course, exchange radiation with space. Perhaps it is through this latter phenomenon that the net radiation will be away from the Earth and the Earth can get rid of its excess energy in this manner. (This will be discussed later.)

To begin to work our way into the subject of atmospheric transmittance, study the details of Figure 7.2. For most purposes, certainly as far as our weather and the absorption of infrared radiation is concerned, the height of the Earth's atmosphere is about five miles. In studying Figure 7.2, imagine that you are on the side of the Earth facing the Sun and that at the top of the Earth's atmosphere there is the complete spectrum of electromagnetic radiation coming from the Sun. The reader should not try to give a spatial interpretation to this figure. You are at one location and far up above you, i.e., at the top of the atmosphere, there is this complete electromagnetic spectrum of radiation headed toward you.

The purpose of Figure 7.2 is to show what happens to this radiation as it comes through the atmosphere toward you on the surface of the Earth. The names of the various regions of the spectrum are indicated at the top of the Earth's atmosphere. Try to appreciate that a linear wavelength scale has not been used because of the great dynamic range of wavelengths involved. The author constantly urges students to try to develop a feeling for just how small a piece of the entire spectrum is the part occupied by the visible. The incoming radiation at the top of the atmosphere is represented by wiggly arrows. It should be noted that some of these wiggly arrows make it through the atmosphere and arrive at the surface of the Earth while others do not get through. Those regions where they do not get through have been blackened. The atom or molecule responsible for absorbing the radiation is indicated on the figure.

Let us direct our attention initially to those molecules responsible for the absorption taking place in the infrared. One immediately notes that it is the minor constituents, namely, water vapor and carbon dioxide. No attempt has been made to show any type of line or band structure associated with these various absorptions. Instead, they are being presented as infinitely sharp absorptions and transmissions.

For wavelengths short of the visible, i.e., ultraviolet, x-rays, and gamma rays, we note that ozone, oxygen, nitrogen, hydrogen, and helium are responsible. Finally, in the radio region beginning at a wavelength of about 30 cm and continuing to longer wavelengths, we note that no radiation reaches the Earth due to reflection by electrons in the ionosphere.

As indicated on Figure 7.2, there are four attenuating mechanisms responsible for radiation not reaching the surface of the Earth, namely,

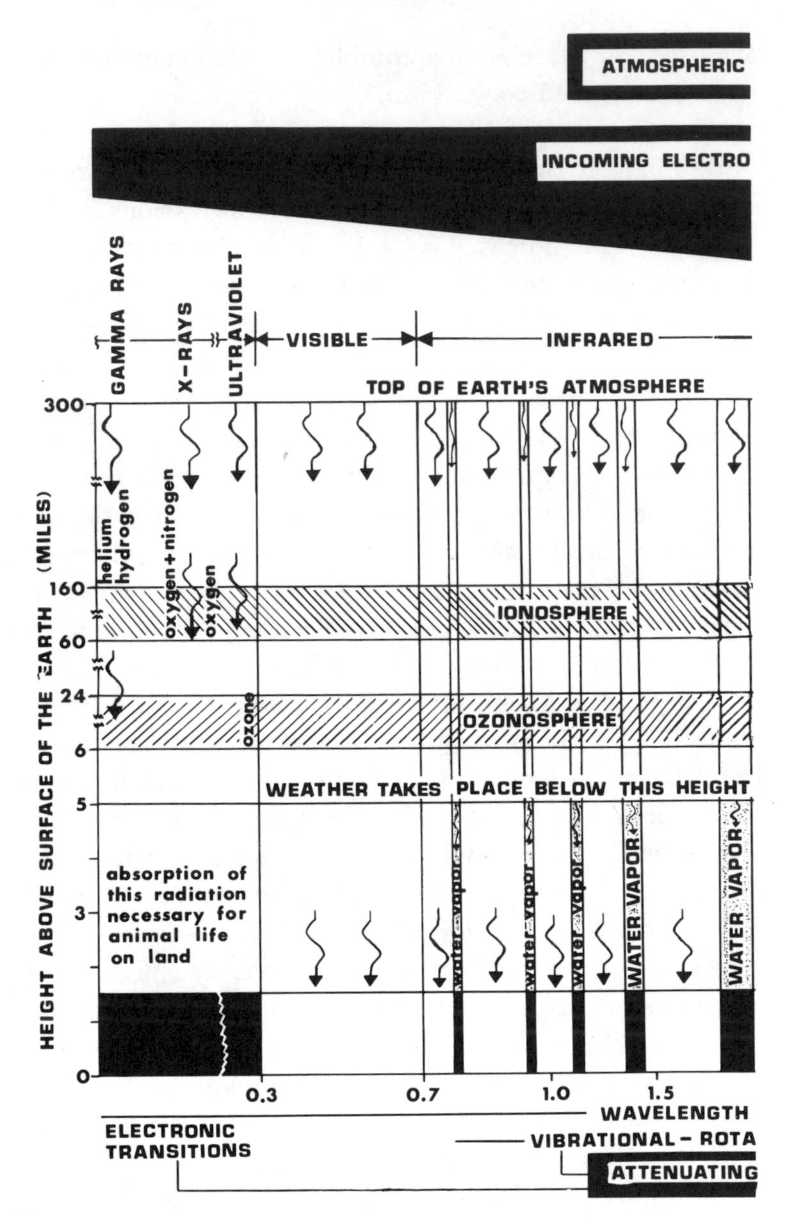

Figure 7.2: Atmospheric attenuation mechanisms.

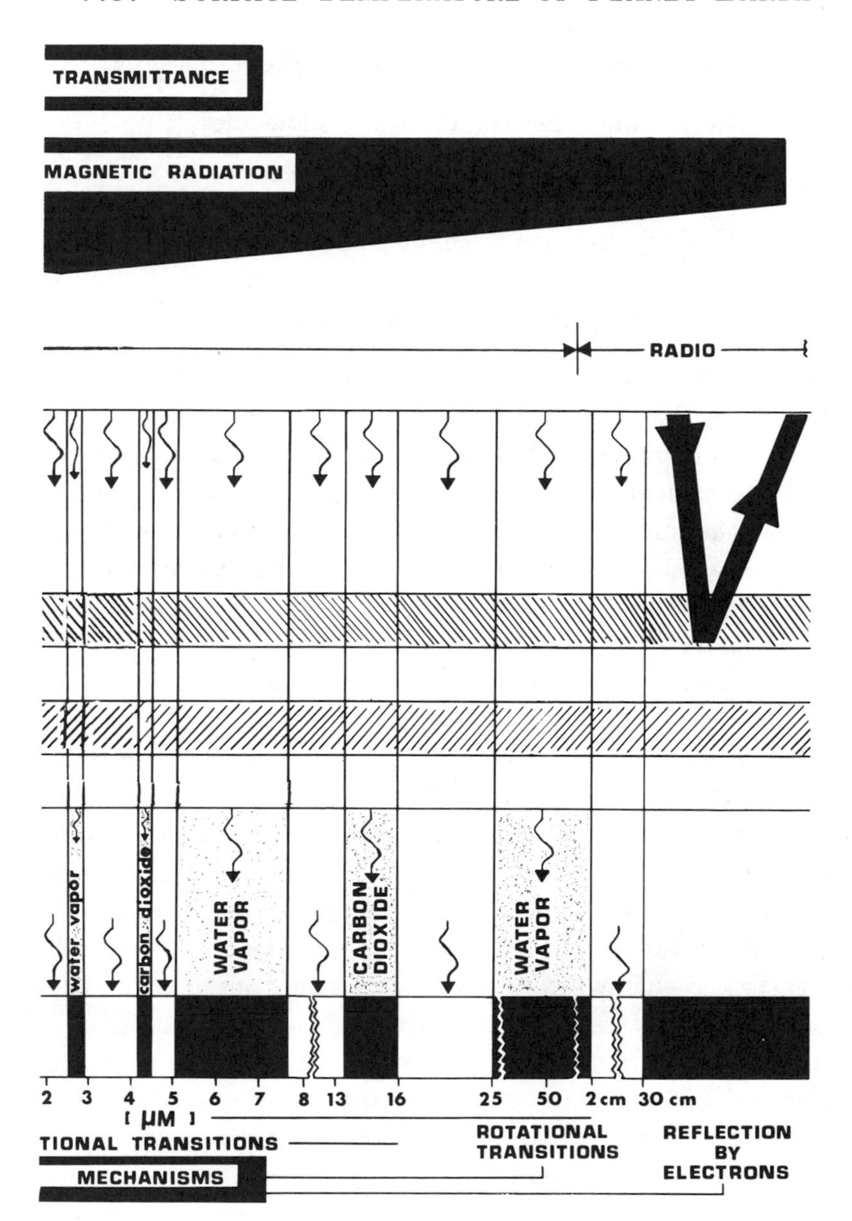
TRANSMITTANCE
MAGNETIC RADIATION
RADIO
water vapor
carbon dioxide
WATER VAPOR
CARBON DIOXIDE
WATER VAPOR
2 3 4 5 6 7 8 13 16 25 50 2 cm 30 cm
[μM]
TIONAL TRANSITIONS
ROTATIONAL TRANSITIONS
REFLECTION BY ELECTRONS
MECHANISMS

- electronic transitions in atoms and molecules,
- vibrational-rotational transitions in molecules,
- rotational transitions in molecules, and
- reflection by electrons in the ionsphere.

Why Study Atmospheric Transmittance?

Everyone should have some familiarity with atmospheric transmittance. In this section we are pursuing it to help us understand the temperature of the Earth. An incident that happened in the Pacific theater of operations during World War II further illustrates the importance of atmospheric transmittance. To improve the spatial resolution of the airborne radar in United States fighter aircraft, someone decided to redesign the radar system and use a shorter wavelength. Theoretically, of course, this is the way to proceed, i.e., by using shorter wavelengths there is less diffraction and one does indeed improve the spatial resolution. (Note: For aerodynamic reasons it was not feasible to increase the size of the radar antenna.) After many months of construction and millions of dollars in expense, the new radar systems were installed. The pilots found that the new systems worked beautifully. They had greatly enhanced spatial resolution. However, for the radar systems to work properly, the pilots had to be so close to the enemy aircraft that they could just as well have used their eyes instead. What had been done in the design of the new radar system was that they had put the frequency (or wavelength) of the radar very nearly on top of one of the lines in the pure rotation spectrum of water vapor. In other words, with the new radar system, they had purchased spatial resolution at the expense of range. Such a blunder was inexcusable! Someone did not take the trouble to find out about atmospheric transmittance.

It will be noted from Figure 7.2 that there are certain regions in the spectrum where radiation gets through to the surface of the

Earth. These regions, referred to as the *atmospheric windows*, are defined between the "centers" of the absorption bands and not between the edges of the windows as follows:

Window Number	Wavelength Limits (μm)
0	0.30 to 0.72
I	0.72 to 0.94
II	0.94 to 1.13
III	1.13 to 1.38
IV	1.38 to 1.90
V	1.90 to 2.70
VI	2.70 to 4.30
VII	4.30 to 6.00
VIII	6.00 to 15.00
IX	15.00 to 25.00

Since water vapor and carbon dioxide do not absorb in the visible, most of the solar radiation is able to penetrate the atmosphere and is absorbed by the Earth. The temperature of the Earth is about 23°C or roughly 300 K. Invoking the Wien Displacement Law, we calculate that the peak radiation from the Earth will take place at about 10 μm. However, this infrared radiation from the Earth is absorbed by water vapor and carbon dioxide. Hence, the atmosphere becomes hotter. As a result, it radiates more to space and more toward the Earth. We thus have a situation that has certain similarities to a greenhouse. One can readily appreciate then that in our present highly technical society, the presence of more carbon dioxide will drive the temperature upward. It should be mentioned, however, that particulate matter in the atmosphere reflects away incoming solar radiation and drives the temperature of the Earth downward. It is difficult to calculate which of these processes is winning and whether or not it will continue to win.

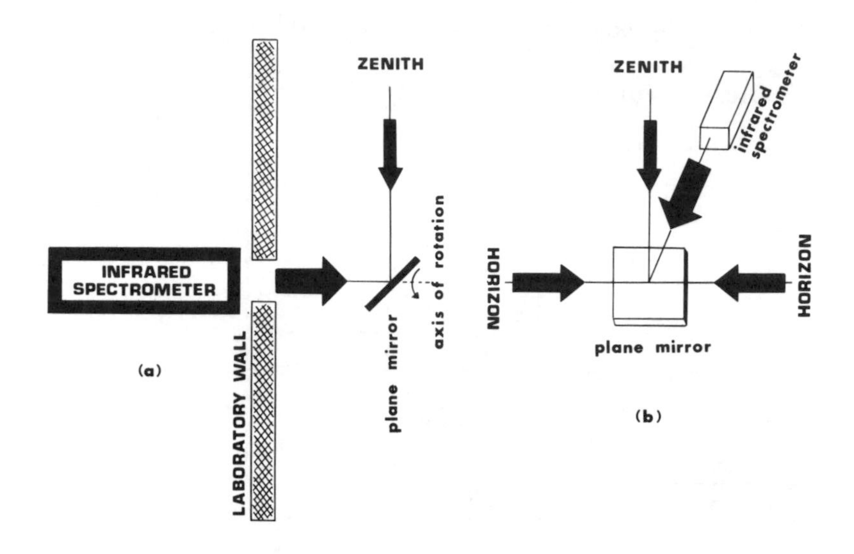

Figure 7.3: Schematic drawing of experimental setup for studying radiation exchange between Earth and space.

Measuring Radiation Exchange between Earth and Space

To carry out this experiment, one would need infrared spectroscopic equipment similar to that sketched in Figure 4.1, plus ancillary optical equipment consisting of a first-surface paraboloidal mirror and one or more plane first-surface mirrors to set up an optical system comparable to that shown in Figure 7.3. One wants to be able to view the zenith sky as well as the horizon sky during both daytime and nighttime hours throughout the spring, summer, fall, and winter. In other words, we want to measure radiation exchange between Earth and space. We do not want to look directly at the Sun.

What one obtains is shown schematically in Figures 7.4(a) and (b). It should be noted that part of the signal obtained is

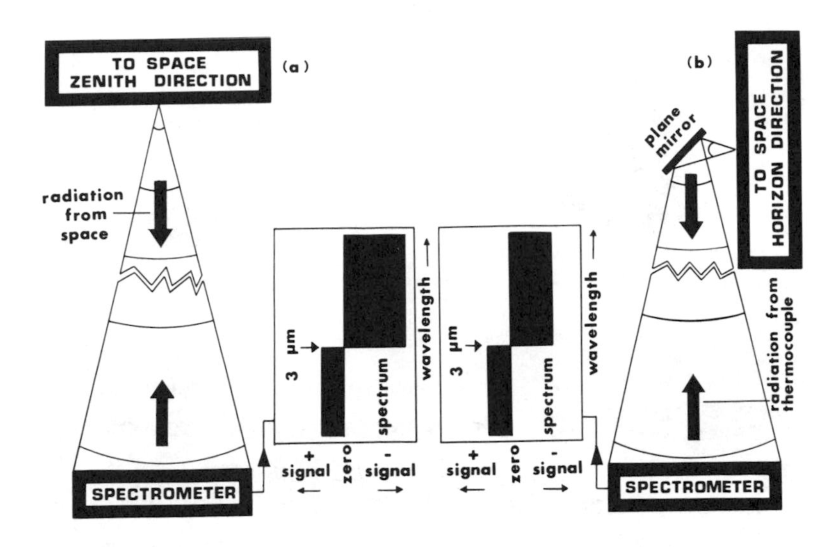

Figure 7.4: Measuring radiation exchange between Earth and space.

positive and part is negative, and that radiation from the zenith sky yields a more negative signal than does radiation from the horizon. One should think about this observation until it makes intuitive sense. The location in the spectrum of the wavelength at which the signal changes from positive to negative is 3 μm, whether one is recording radiation from the zenith or from the horizon.

Interpretation of Radiation Exchange Experiments between Earth and Space

Figure 7.5 summarizes the results of radiation exchange experiments between Earth and space. Figure 7.5(a) indicates the situation on the sunlit surface of the Earth. In other words, the Earth receives radiation from the Sun in the visible and gets rid of energy by radiating in the 8 to 12 μm region. Figure 7.5(b) indicates the situation on the non-sunlit surface of the Earth.

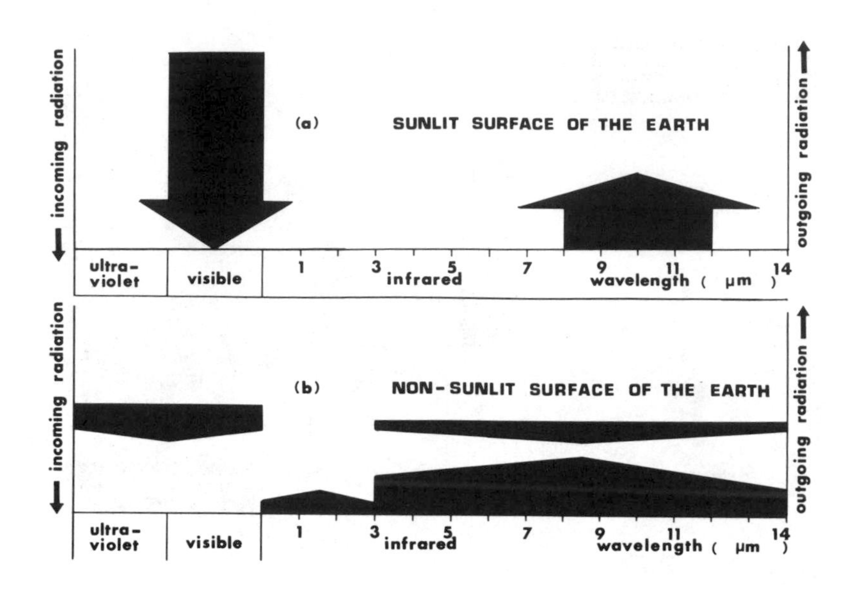

Figure 7.5: Radiation exchange between the sunlit surface of the Earth and space and between the non-sunlit surface of the Earth and space.

Figure 7.6 represents an interpretation of radiation exchange experiments between Earth and space, i.e., not looking directly at the Sun.

The author's first interest in experiments dealing with the exchange of radiation between Earth and space was due to the report, "On the Exchange of Radiant Energy between the Earth and the Sky," by Clay P. Butler (Naval Research Laboratory Report 3984, June 11, 1952).

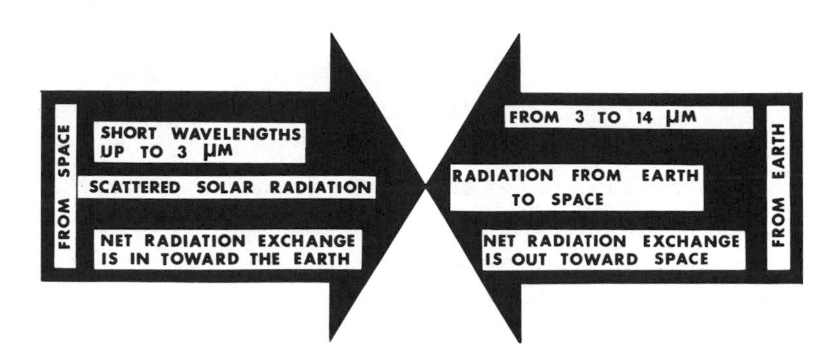

Figure 7.6: Interpretation of radiation exchange experiments between Earth and space.

7.4 Decoding Messages from Atoms and Molecules

Perhaps the reader has not given much thought to the fact that atoms and molecules are constantly communicating with us. It will be quite interesting to investigate what types of information we might learn if we could be clever enough to decode their messages.

Many of the things that have already been discussed in this book will be of help in this decoding challenge. For example, when one locates a radiometer in the focal plane of a telescope and points it toward empty space in the skies, will one be clever enough to decode the messages contained in what is called the *3 K background radiation*? Remote sampling experiments provide an excellent opportunity for us to learn that atoms and molecules identify themselves with their *optical signature* or *spectrum*. The science of using a spectrometer (or a radiometer) to read and decode messages from atoms and molecules is called *spectroscopy*. It should be pointed out that is was Kirchhoff, Professor of Physics at Heidelberg University, who was the first to realize that the spectrum of an atom is unique, i.e., that the spectrum from each atom is different.

In addition to identifying themselves, atoms and molecules also tell us many other things about themselves and their environment. For example, they tell us whether they are moving toward us or away from us and at what velocities. Even the presence or absence of magnetic and electric fields is included in these communications. Spectroscopy has had a profound influence on the development of physics, particularly quantum physics.

7.4.1 Remote Sampling

Decoding the messages from distant atoms and molecules is referred to as ***remote sampling*** or ***remote sensing***. If one is carrying out remote sampling experiments from the surface of the Earth or through the Earth's atmosphere from space, say, it is necessary to know about how the Earth's atmosphere can modify the electromagnetic radiation coming from the atoms and molecules. Obviously, such measurements would have to be carried out in the "atmospheric windows."

When one begins to think about carrying out remote sampling experiments, one will realize there are only two ways of doing this—study the self-radiation (often referred to as thermal radiation) emitted by the object (the "passive" approach) or reflect (or scatter) radiation from the object (the "active" approach). Figure 7.7 illustrates these two methods. From these drawings, one can see very clearly the importance of the part played by the Earth's atmosphere. The following listing should help one to understand this figure.

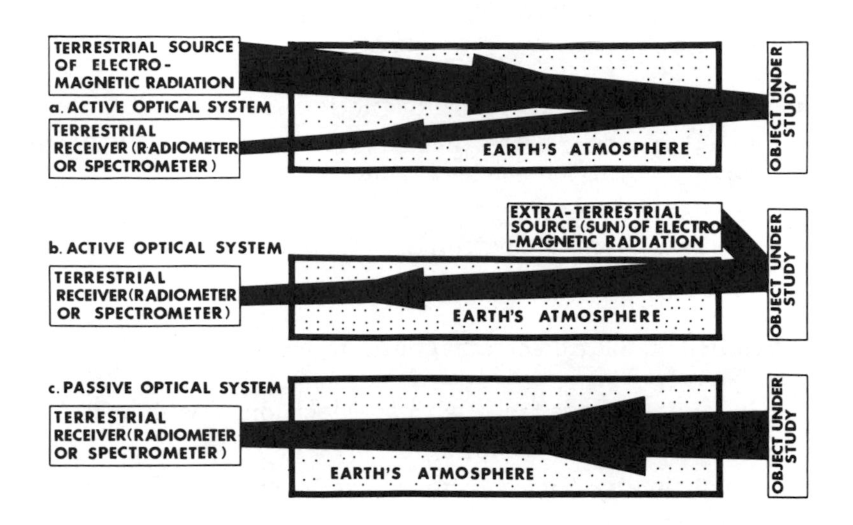

Figure 7.7: Active versus passive optical systems for carrying out remote sensing (or remote sampling).

Active Optical System

Figure 7.7(a):

- Radar studies of planets or some object in the Earth's atmosphere,
- Use of lasers to study atmospheric pollution,
- Radar and laser ranging

Figure 7.7(b):

- Planetary studies with the Sun as the source of reflected (or scattered) electromagnetic radiation

Passive Optical System

Figure 7.7(c):

- Studies of thermal electromagnetic radiation from objects located in the Earth's atmosphere or outside the atmosphere,
- Studies of thermal electromagnetic radiation from the planets,
- Studies of solar electromagnetic radiation

At this point, one is faced with the decision of which example or examples of remote sampling to present to students. There are obviously many topics from which to choose. It has been the author's experience that students are going to be interested in your choice, regardless of what it is. They just seem to be interested in this type of physics.

Suggestions that come immediately to mind would be discussions of some of the very interesting observations made from man-made satellites of the distribution of crops on the planet as well as meteorological observations. One might also want to discuss some of the schemes that have been used to try to detect "clear air turbulence" in the atmosphere. The example below deals with planetary radiometry and spectroscopy.

7.4.2 Planetary Radiometry and Spectroscopy

It is important to be sure that one fully understands the origin of planetary radiation. Figure 7.8 is a schematic representation of the four ways in which radiation can reach the Earth from a planet. In the visible and near infrared, scattered solar radiation from the planet's surface is dominant. Figure 7.9 is an attempt to illustrate the origin of a planet's spectrum. In this figure, *thermal emission* is what we have previously called *self-radiation.*

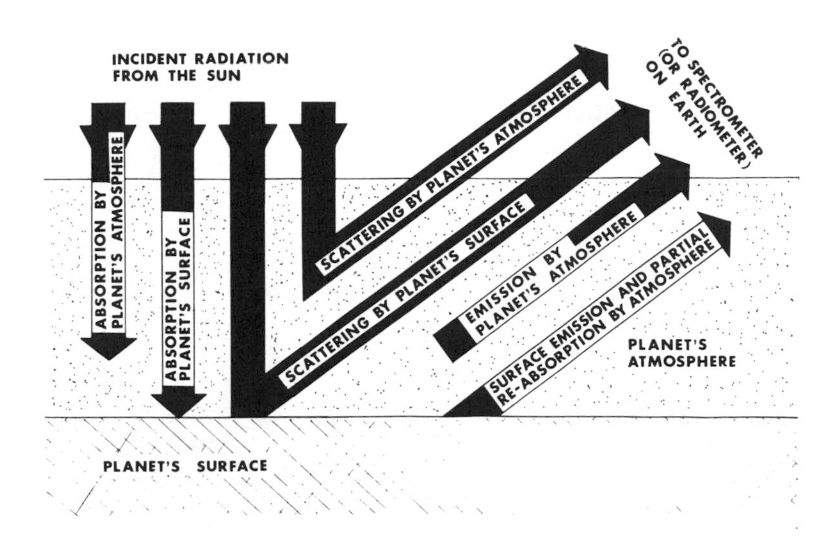

Figure 7.8: Schematic representation of the four ways in which radiation can reach the Earth from a planet. In the visible and near infrared, scattered solar radiation from the planet's surface is dominant.

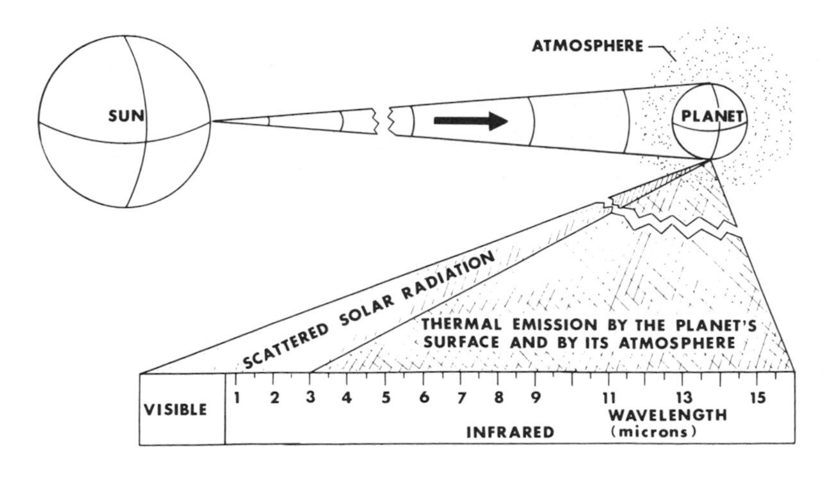

Figure 7.9: Origin of a planet's spectrum.

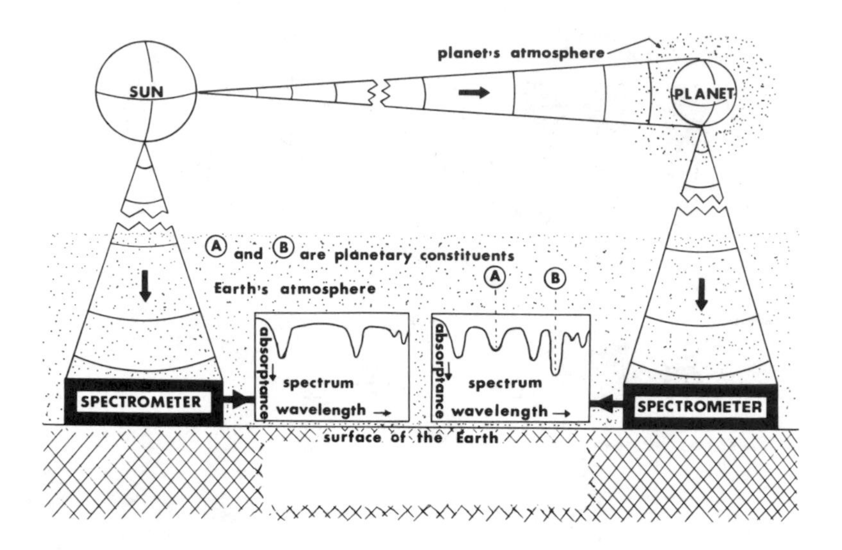

Figure 7.10: Determining the constituents in a planet's atmosphere.

Figure 7.10 illustrates what is involved in determining the constituents in a planet's atmosphere. Basically, one looks at the Sun with a spectrometer and records its spectrum. Next, one looks at the planet with a spectrometer and records its spectrum. One then looks for differences, such as (A) and (B) in Figure 7.10. One then has to determine in the laboratory what gas, or gases, absorb in the regions indicated by (A) and (B).

For those readers who would like to pursue the study of planetary radiometry and spectroscopy in greater detail, there exists a very extensive literature, both of texts and published articles. To assist you in this matter the following listing of texts and articles is given for your convenience.

1. *Astrophysical Techniques*, C. R. Kitchin (Adam Hilger, 1984).

2. *Astronomical Spectroscopy*, A. D. Thackeray (Macmillan, 1961).

3. *The Earth and its Atmosphere*, D. R. Bates, ed. (Basic 1957).

4. *The Atmospheres of the Earth and Planets*, G. P. Kuiper, ed., (University of Chicago, 1949).

5. "Infrared Spectroscopy of Planets and Stars," W. M. Sinton, *Applied Optics*, March 1962, 105.

6. "Recent Infrared Spectra of Mars and Venus," W. M. Sinton, *Journal of Quantitative Spectroscopy and Radiative Transfer*, **3**, 4, 1963.

7. *Space Science Reviews*, "Molecular Spectroscopy of Planetary Atmospheres," **1**, 1, June 1962, 159.

8. "Planetary Atmospheres," *The Physics Teacher*, May 1973, 277.

Part III

Quantification of Electromagnetic Radiation

Chapter 8

Radiometric Calibration

8.1 Introduction

There are three basic problems in either spectroscopy or radiometry, namely, (a) measurement of wavelengths, (b) measurement of intensities, and (c) interpretation. This chapter is about intensity measurements, i.e., "how much" radiation there is. In particular, this chapter deals with the radiometric calibration of radiometers, spectrometers, and thermopiles.

The chief purpose of this chapter is to outline certain fundamental concepts and methods of radiometry and spectroscopy adequately so that they can serve as a guide to further study. The author is concerned that many people, for one reason or another, never really learn how to measure a "watt" of electromagnetic radiation. Unless one is careful in this matter, it is quite likely that one will never really learn how to quantify radiation. Most of the undergraduate laboratory manuals, particularly those dealing with optical physics, are completely lacking in this area.

Radiometry as discussed herein relates to the measurement of radiant energy in any part of the electromagnetic spectrum. If the radiant energy being measured lies in the visible part of the electromagnetic spectrum, it is referred to as *photometry*.

The author likes to approach radiometry with vigor rather than with rigor. For example, the wavelength calibration of spectrometers is pursued only to the point of giving modest results in the measurement of wavelengths, and he likes to convey to the reader some of the very interesting possibilities in this area of physics.

Most of the emphasis in this chapter is on infrared radiation. It is in the infrared region of the electromagnetic spectrum that so many physical phenomena can be interestingly demonstrated, and also it is the spectral region where the radiation laws can be pedagogically presented.

It is much more difficult to measure intensities accurately than it is to measure wavelengths accurately. The state of the art is such that wavelengths can be measured to about 1 part in 10^8 and intensities to about 1 part in 10^2. Measurements made in the infrared and longer wavelengths are particularly difficult. Operation in this region of the spectrum has been compared with carrying out measurements in a sea of radiation where all objects are radiating and exchanging energy with each other. Hence, the experimenter has to be extremely careful. Activity in this area of physics forces one to become quantitative in thinking about radiation. Without this type of experience, many students never really learn how to quantify radiation.

8.2 Infrared Prism Spectrometer

Let us assume that our spectrometer to be radiometrically calibrated is a Perkin-Elmer model 12 C infrared spectrometer. As a dispersing element, let us assume we are using a sodium chloride (NaCl) prism and that our detector is a thermocouple.

8.2.1 Wavelength Calibration

Since this chapter is primarily about the radiometric calibration of this spectrometer, we also must assume that the spectrometer has been wavelength calibrated. However, we need to say a few things about its calibration. In calibrating spectrometers for use in the visible and ultraviolet, one primarily uses the techniques of emission spectroscopy. In the very near infrared, say to about 1.5 μm, one also uses emission techniques. As one goes to longer wavelengths in the infrared (say beyond 1.5 μm), the wavelength calibration is carried out by using absorption techniques. If one is involved in the wavelength calibration of a commercially manufactured spectrometer, the first thing to do is to consult the instruction manual accompanying the spectrometer.

Various gases or vapors, such as mercury, sodium, potassium, helium, neon, thallium, xenon, zinc, cadmium, rubidium, cesium, iodine, bromine, argon, krypton, water vapor, and nitrogen, should be readily available in most laboratories along with Osram lamps, gas discharge tubes, etc. The emission lines from these gases and vapors will be quite useful in wavelength calibration. To use these emissions sources, it is necessary to drill and tap two holes in the base of the spectrometer so that the radiation chopper can be located in front of the entrance slit rather than in front of the globar radiation source as the instrument is normally used.

An accessory item that accompanies this spectrometer is a 10 cm long absorption cell (about 5 cm in diameter). This absorption cell is used to obtain the absorption spectra for such gases and vapors as CO_2, CO, NH_3, HCl, and HBr. The manufacturer's instruction manual provides these calibration spectra. The article by Plyler *et al.* [1] tabulates suitable bands of common gases, remeasured wherever necessary, from 2 to 16 μm to obtain an accuracy of about 0.03 cm^{-1} throughout the region and to provide good calibrating points at frequent intervals. The substances used are the following: H_2O, CO_2, CO, HCl, HBr, NH_3, C_2H_2, CH_4, N_2O, and polystyrene film [2,3].

In carrying out wavelength calibration, one needs to be aware of the importance of interference phenomena, in particular, the so-called Edser-Butler technique. These Edser-Butler bands are

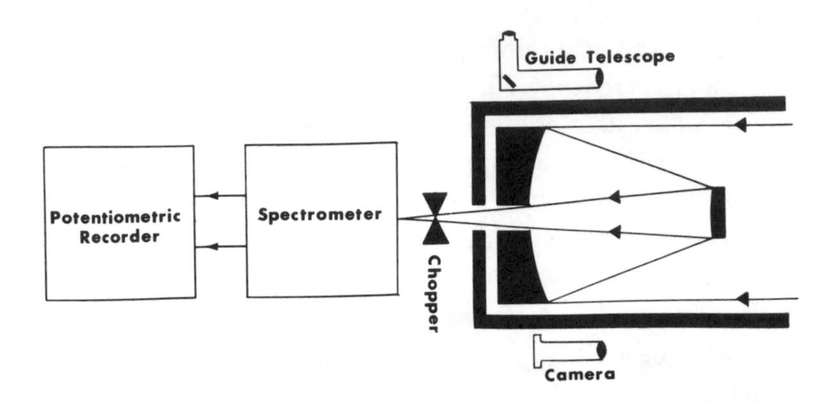

Figure 8.1: Schematic sketch indicating the use of a radiometrically calibrated spectrometer being used with a Cassegrain-type telescope.

of great help in measuring the wavelength of unknown lines located between known wavelengths. If one is interested in pursuing this phenomenon, the article listed in reference 3 at the end of this chapter is highly recommended.

8.2.2 Radiometric Calibration: Theory Involved

To get a feeling for what we are going to do in this chapter, consider Figure 8.1. Let us assume the read-out device is a potentiometric recorder. If the optical system sketched in Figure 8.1 is pointed at a planet, say, we are going to get a reading (i.e., a pen deflection) on the recorder. What we would like to be able to do is to measure the recorder deflection and, knowing the electronic gain setting of the amplifier used with this system, somehow be able to say that the pen deflection corresponds to a certain amount of radiation (i.e., a certain power level expressed in watts striking the detector).

It should be noted that a potentiometric recorder produces a

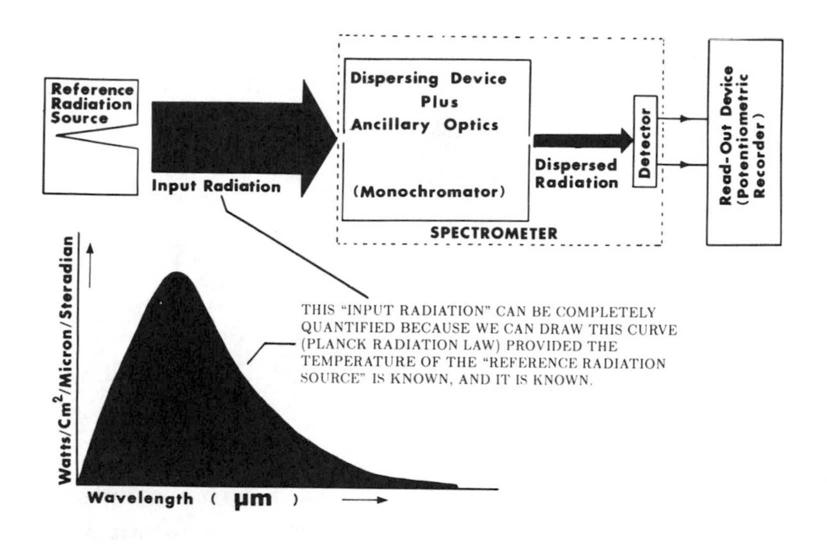

Figure 8.2: Pictorial representation of some of the key ideas involved in radiometric calibration.

deflection on the paper that is proportional to the voltage appearing at the input terminals of the recorder. Recorders, obviously, vary in their full-scale sensitivities. We have found useful the recorders (Leeds & Northrup Type G) that require a 10 millivolt (direct current) signal for full-scale deflection. This means, of course, that a recorder deflection of half of full scale would correspond to an input of 5 millivolts. Therefore, any radiation coming into the spectrometer will produce a certain signal level on the recorder paper that can be measured and expressed in millivolts.

Let S equal the deflection on the recorder (measured in millivolts, mV) produced by a beam of radiation falling on the detector. What would be desirable is to know by what factor (say K) the deflection, S, needs to be multiplied to yield the number

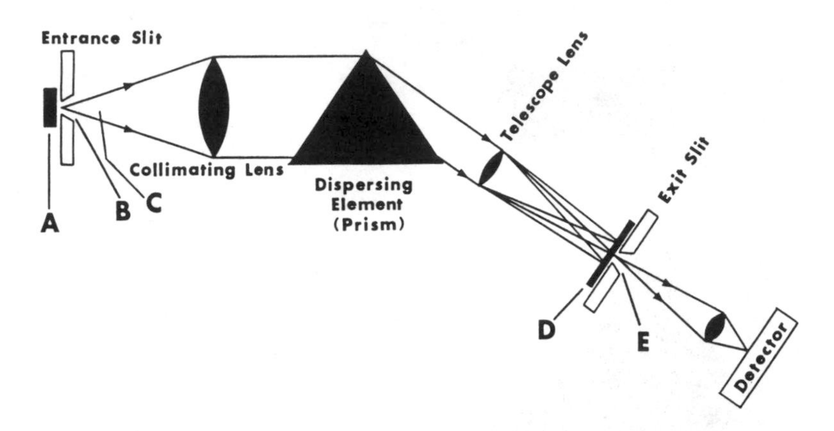

Figure 8.3: Sketch showing additional ideas involved in radiometric calibration. (A) Either a real image of the *reference radiation source* or the *reference radiation source* itself. (B) The area of this entrance slit determines the area of the *reference radiation source* that radiates into the spectrometer. (C) This solid angle determines the solid angle over which radiation enters the spectrometer. (D) The real image of spectrum is formed on this plane. (E) Part of the radiometric (or absolute) calibration involves calculating the amount of power (in watts) passing through this exit slit. This is a theoretical calculation and all losses are neglected.

of watts falling on the detector. This calibration factor, K, will have the following units:

$$K = \text{calibration factor (watts/millivolt)}.$$

If we let P equal the power falling upon the detector, we can write

$$P = K\,\text{(watts/millivolt)}\;S\,\text{(millivolts)} = K\,S\,\text{(watts)}.$$

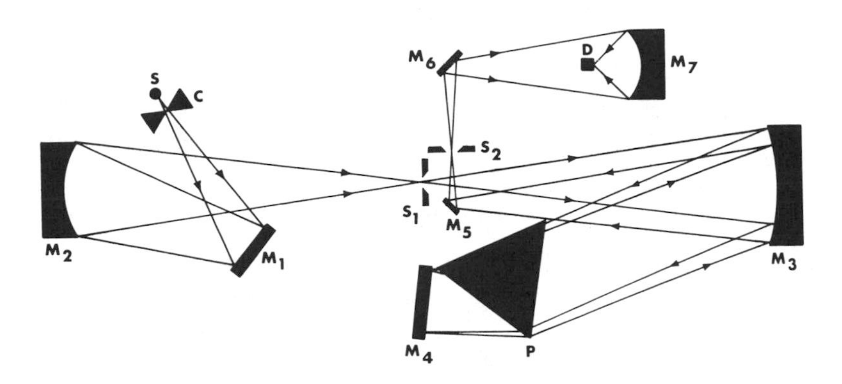

Figure 8.4: Perkin-Elmer model 12 C infrared prism spectrometer (Littrow arrangement). S = globar source, C = radiation chopper, M_1, M_4, M_5, M_6 = plane mirrors, M_2 = spherical mirror, M_3 = paraboloidal mirror, M_7 = ellipsoidal mirror, D = thermocouple, P = sodium chloride prism.

To carry out the determination of this calibration factor, K, it will be necessary to deal with radiation whose spectral characteristics can be calculated theoretically. One will recall that Planck's Radiation Law (Chapter 3) enables us to calculate the distribution of radiation from a blackbody provided we know the temperature of the blackbody. Figure 8.2 will help to clarify what our procedure will be. The *reference radiation source* shown in Figure 8.2 is either a commercially available blackbody or a homemade blackbody. Further details of our procedure are sketched in Figure 8.3.

Let us begin our theoretical discussion by considering the infrared spectrometer shown in Figure 8.4. Let us assume that all optics to the left of the entrance slit, S_1, have been removed (i.e., M_1, M_2, C, and S). We then have what is called a *monochromator.* Imagine that we have a reference radiation source located just outside the entrance slit of the monochromator.

Consider the simplified spectrometer shown in Figures 8.5(a) and (b). Let

s = width of the entrance and exit slits (Note: In the Perkin-Elmer spectrometers they open and close together.);
L = length of entrance and exit slits;
A = aperture of collimating lens and telescope lens;
f = focal length of collimating lens and telescope lens;
P = power intercepted by the collimating lens.

The expression for P is given by the following:

$$P = \frac{1}{\pi}\int_0^\infty \frac{c_1}{\lambda^5}\frac{1}{\left[e^{\frac{c_2}{\lambda T}} - 1\right]}\, d\lambda \frac{\text{watts}(sL)\text{cm}^2}{\text{cm}^2\text{ steradian}}\left[\frac{A}{f^2}\right]\text{steradian} \qquad (8.1)$$

where

c_1 = first radiation constant = 3.732×10^{-12} watt cm^2 = $2\pi c^2 h$;
c_2 = second radiation constant = 1.436 cm deg = hc/k. (Note: h = Planck's constant; c = velocity of light; k = Boltzmann constant.);
sL = area of entrance slit;
A/f^2 = solid angle subtended by the collimating lens at the entrance slit.

If we let $W(\lambda, T)$ equal the *spectral radiant emittance* of the source, then Equation (8.1) can be written as follows:

$$P = \frac{1}{\pi}\int_0^\infty W(\lambda, T)\, d\lambda\, (sL)\, (A/f^2) \text{ watts.} \qquad (8.2)$$

In this development we shall not try to account for absorption, scattering, and reflection loss in the system because these losses are inherently involved in the measurement of the calibration factor (or sensitivity factor), K.

To complete our radiometric calibration, we need to know the *spectral slit width* of the spectrometer. By this we mean the *wavelength interval embraced by the exit slit.* Figure 8.5(b) will assist in visualizing the concept of spectral slit width. From this figure one notes that

$$\begin{aligned}
\Delta\theta &= s/f; \\
\Delta\lambda &= \text{wavelength interval embraced by the exit slit,} \\
\Delta\lambda &= \frac{\Delta\theta}{\left(\frac{d\theta}{d\lambda}\right)} = \frac{s}{f\left(\frac{d\theta}{d\lambda}\right)}.
\end{aligned}$$

Note also that

$$\left(\frac{d\theta}{d\lambda}\right) = \left(\frac{d\theta}{dn}\right)\left(\frac{dn}{d\lambda}\right),$$

where n = index of refraction of the prism.

8.2.3 Radiometric Calibration: Step-By-Step Procedures Involved

In carrying out this radiometric calibration, the following procedure is suggested:

1. Set the temperature of the reference radiation source. Once the temperature has been set, one can calculate $W(\lambda)$ by using the Planck Radiation Law or else by using some of the published tables giving these data in the literature [4,5].

2. Set the width of the entrance slit (and hence exit slit) to give an easily measured deflection on the recorder.

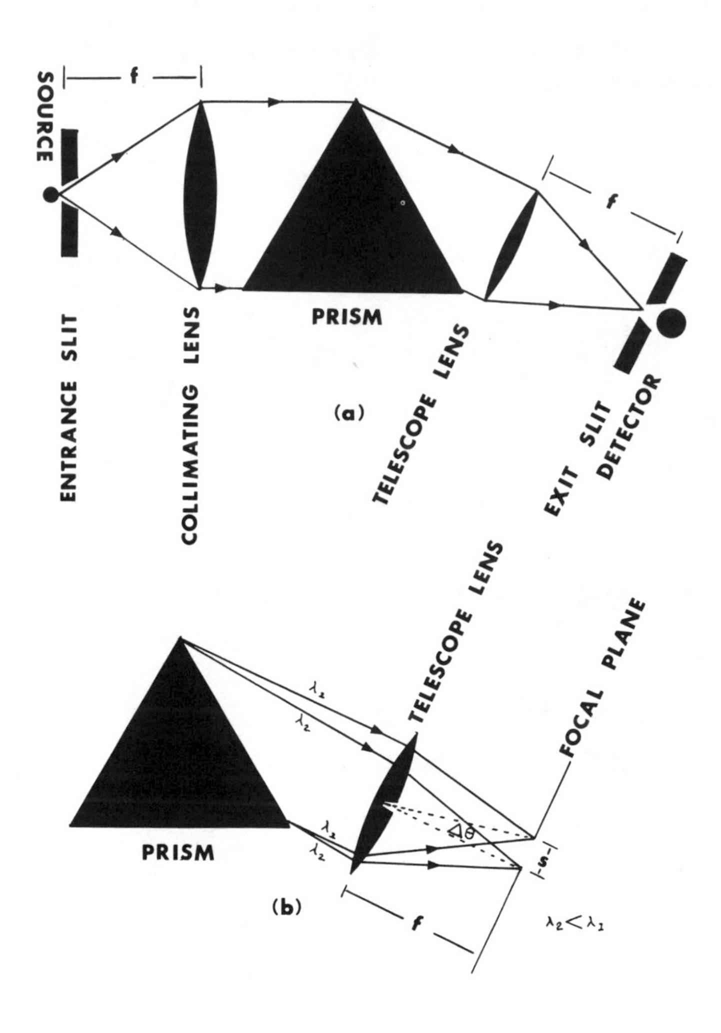

Figure 8.5: A simplified prism spectrometer.

3. Put on the recorder an input signal of known direct current voltage (in millivolts or microvolts).

4. Calculate the spectral slit width from knowing the mechanical width of the exit slit, $dn/d\lambda$ of the prism and the focal length of the telescope lens.

5. Make this spectral slit width calculation for various wavelengths throughout the spectral region to be covered by the particular prism and detector.

6. From a knowledge of the spectral slit width, calculate the power striking the detector.

7. From a knowledge of the power striking the detector at the various wavelengths and from the various recorder deflections produced, calculate the calibration factor (or sensitivity factor), K, at various wavelengths and make a graph of K versus λ.

Let $(\lambda_2 - \lambda_1) = \Delta\lambda =$ spectral slit width (or spectral bandpass). For purposes of simplification we shall assume that

$$\int_{\lambda_1}^{\lambda_2} W(\lambda)\, d\lambda \approx W(\lambda)\, \Delta\lambda,$$

where $W(\lambda)$ is evaluated at the midpoint of the wavelength interval, λ_1 to λ_2, and is considered constant over the interval $\Delta\lambda$.

Suppose, for example, that one wants to calculate the power striking the detector at 10 μm. One could either calculate $W(\lambda)$ at the known temperature of the source using Planck's Radiation Law or look up the value in some appropriate table (such as the tables by Pivovansky and Nagel). One next multiplies by $\Delta\lambda$ at 10 μm for the particular prism and spectrometer involved.

We now have a spectrometer that is radiometrically calibrated in case we choose to locate a source immediately in front of the

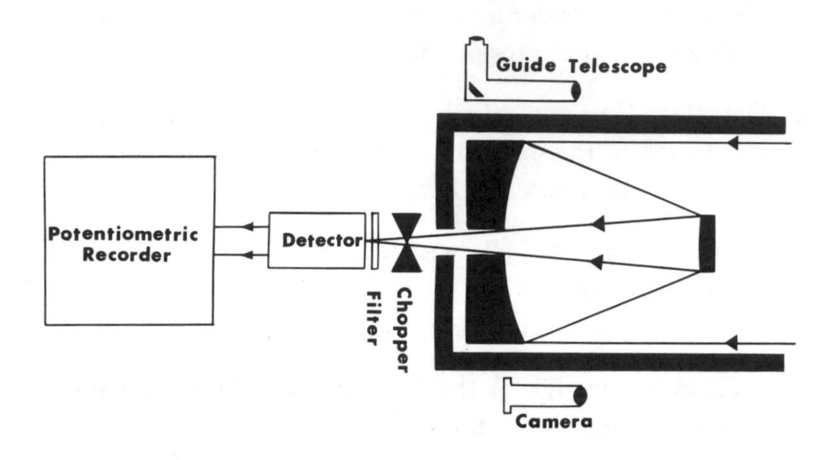

Figure 8.6: Basic components of a radiometer (Cassegrain type).

entrance slit. Suppose, however, that we become interested in something other than this, say, for example, in the radiation from a distant planet collected by a telescope and fed into the entrance slit of the spectrometer, as shown in Figure 8.1 or as sketched in Figure 8.6 in the case of a radiometer. The obvious complication that has been introduced by the systems shown in Figures 8.1 and 8.6 are *two additional reflections.*

Most mirrors used in spectroscopy and radiometry are coated with aluminum. As we have noted, the radiant reflectance of a mirror increases as one goes to longer wavelengths. For certain applications in the far infrared, mirrors are coated with gold because gold has a greater radiant reflectance in these regions than does aluminum.

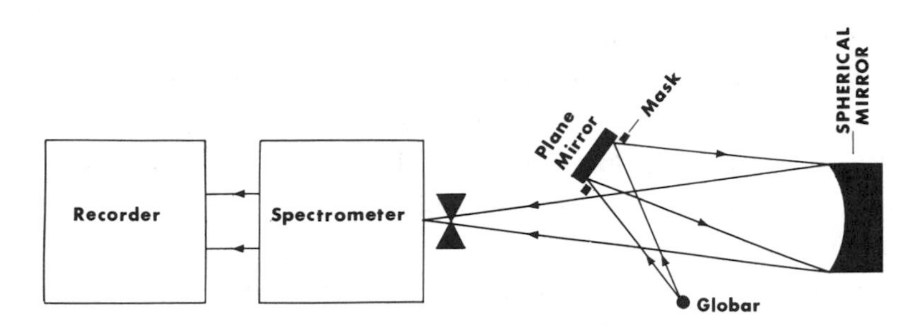

Figure 8.7: Experimental setup for measuring the radiant reflectance of a piece of glass.

8.3 Measuring the Radiant Reflectance of a Dielectric

Let us pause briefly in our theoretical discussion to examine how one might measure radiant reflectance experimentally. For our first experiment, suppose we want to measure the radiant reflectance of a piece of dielectric (say, a piece of glass) for a particular angle of incidence of the radiation. Consider the experimental arrangement shown in Figure 8.7. Let us assume that we want the spectral radiant reflectance of the glass measured in the 0.5 to 15 μm region. This type of experiment could very easily be performed using a sodium chloride (NaCl) prism, a thermocouple detector, and a globar source of continuous radiation. The dynamic range of wavelengths to be covered (5 octaves) is so great that within this range of wavelengths the spectral radiant emittance of the globar will vary greatly. What this means is that to get sufficient deflection on the recorder to be able to measure it accurately, one will have to change, several times, the amplifier gain settings and the width of the entrance and exit slits.

It is imperative that one notes the amplifier and slit width changes that are necessary (i.e., record them on the chart paper when they are made) because it will be necessary to use these data again shortly. Starting first with the plane aluminized mirror located behind the mask (note: the purpose of the mask is to set accurately the area of the reflecting surface being measured), one would obtain a hypothetical curve similar to curve (a) in Figure 8.8(a). One should adjust the amplifier gain and the slit opening to give as large a recorder deflection in the case of aluminum as possible. Next, one carefully substitutes the piece of glass for the aluminized mirror (being very careful not to rotate the mount holding the mask) and repeats the experiment, being sure to use for the piece of glass the same amplifier gain settings and slit settings that were used with the aluminized mirror for each of the wavelength regions scanned. One would obtain a hypothetical curve similar to curve (b) shown in Figure 8.8(a). One then takes the two curves shown in Figure 8.8(a) and divides the recorder deflections obtained from curve (b) by those for curve (a) at various wavelengths throughout the region of interest and finally obtains the radiant reflectance curve for glass, as shown in Figure 8.8(b).

This method for measuring the radiant reflectance of a dielectric is sufficiently accurate for most experiments. Our procedure is equivalent to saying the radiant reflectance of a freshly deposited film of aluminum is 100%!

8.4 Measuring the Radiant Reflectance of Aluminum

For our second experimental problem, let us complicate the situation by asking how to measure the radiant reflectance of alu-

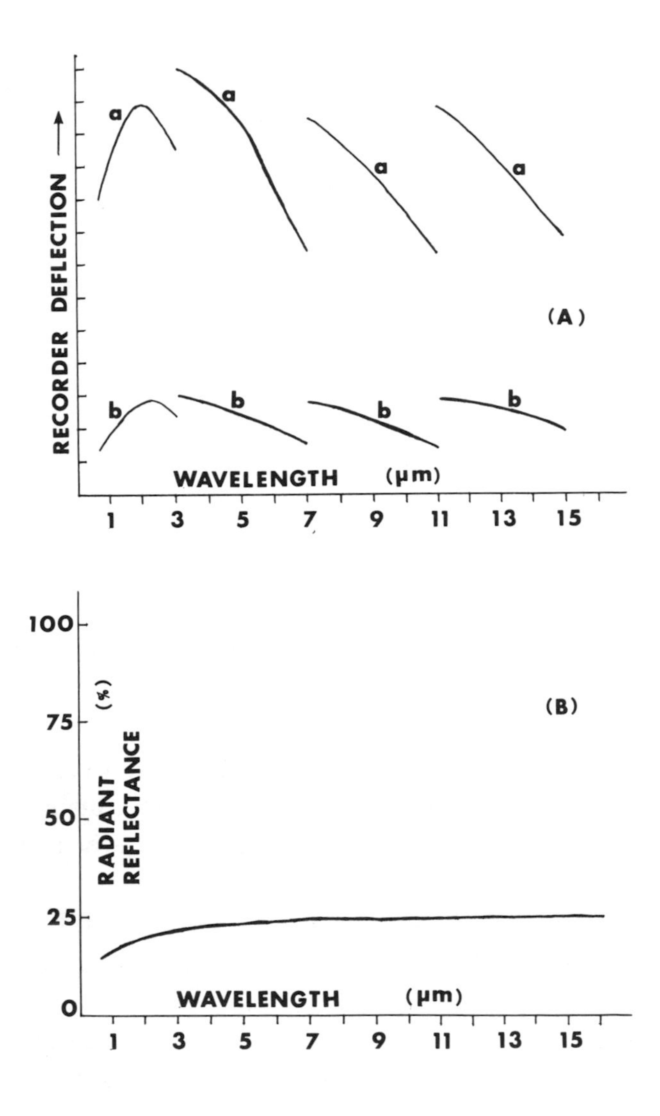

Figure 8.8: Hypothetical radiant reflectance data for glass.

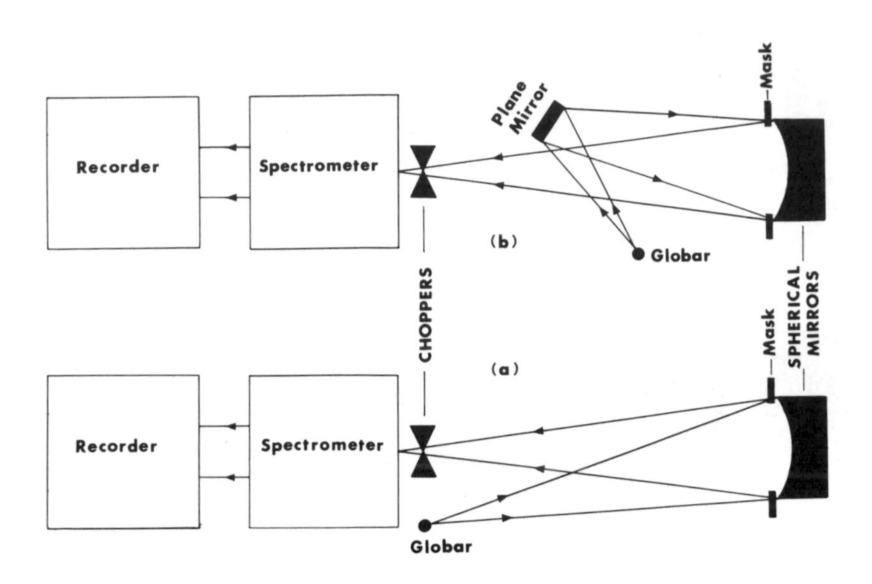

Figure 8.9: Experimental setup for measuring the radiant reflectance of an aluminized mirror.

minum. Because we have no material with which we can compare aluminum, we must resort to an experimental scheme different from that used in Figure 8.7. One possibility is the scheme shown in Figure 8.9.

In this method, one scans the spectral range of interest using the setup shown in Figure 8.9(a) and gets a series of recorder deflections (curve a), as shown in Figure 8.8(a). One then uses the set-up shown in Figure 8.9(b) and gets a series of recorder deflections (curve b), as shown in Figure 8.8(a). From these two curves one can again obtain the radiant reflectance using the same technique that led to Figure 8.8(b). It is imperative in this scheme that the optical path followed by the radiation be the same in both Figures 8.9(a) and (b).

We have just seen how to obtain the radiant reflectance of

a plane aluminized mirror. Referring back to Figure 8.1 and Figure 8.6, it will be seen that what we really want is the radiant reflectance of a paraboloidal mirror and a hyperboloidal mirror. One possible way around this difficulty is the following. When the paraboloidal mirror (or hyperboloidal mirror) is placed in the bell jar used for its aluminizing, be sure to include alongside it a plane piece of glass to be aluminized at the same time. One then measures the radiant reflectance of the plane aluminized piece of glass. In this method, at least the age of the aluminum coating of the plane mirror and the paraboloidal, or hyperboloidal, mirror will be the same. For most purposes this procedure will suffice. For those very few cases where it will not suffice, one must resort to more elaborate schemes that we shall not pursue here.

8.5 Use of Two Collimators to Carry Out Radiometric Calibration

Let us next consider the optical systems shown in Figure 8.10. A moment's contemplation will convince the reader that the use of two collimators is a convenient method of transferring a real image from the collimator on the left to the collimator on the right and conversely. The lower part of this figure will help to make this clear where lenses, instead of mirrors, are shown for pedagogical reasons. The reader should see that the reflective and refractive collimators shown in Figure 8.10 are equivalent regardless of the separation of the collimators. This statement is true, of course, only if one can neglect any atmospheric attenuation in the optical path between the two collimators.

Total power intercepted by mirror $A_1 =$

$$\frac{\sigma T^4}{\pi}\text{watts cm}^{-2}\text{ steradian}^{-1}(a_1)\text{cm}^2\left[\frac{A_1}{f_1^2}\right]\text{steradian}$$

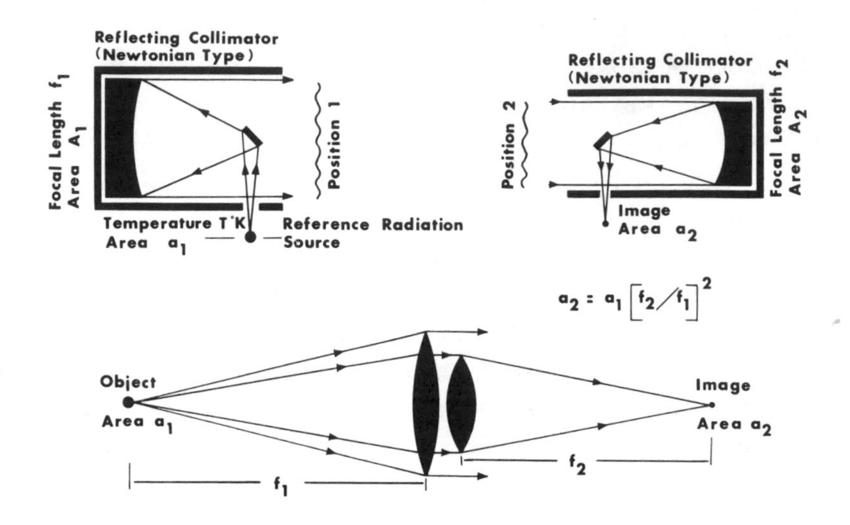

Figure 8.10: Reflecting and refracting collimators [linear magnification = image distance/object distance = f_2/f_1; area magnification = $(f_2/f_1)^2$].

$$= \frac{\sigma T^4}{\pi} a_1 \frac{A_1}{f_1^2} \text{watts.} \tag{8.3}$$

Let R_1 = radiant reflectance of each mirror in the collimator on the left.

Total power density at position 1 =

$$\frac{\sigma T^4}{\pi} a_1 \frac{A_1}{f_1^2} \text{ watts} \, (R_1)(R_1) \, \frac{1}{A_1 \text{cm}^2.}$$

$$= \frac{\sigma T^4}{\pi} \frac{a_1}{f_1^2} R_1^2 \, \text{watts/cm}^2 \tag{8.4}$$

This power density is now incident at position 2, i.e., the collimator on the right. Suppose we next calculate the power (in watts) intercepted by the mirror of area A_2 and focused in the focal plane into an image whose area is given by the following:

Area of image formed in focal plane of collimator on the right $= a_2 = a_1(f_2/f_1)^2$.

Let R_2 equal the radiant reflectance of each mirror in collimator on the right. The total power incident on focal plane in collimator on the right equals

$$\frac{\sigma T^4}{\pi}\frac{a_1}{f_1^2}R_1^2\,\frac{\text{watts}}{\text{cm}^2}\,A_2\ \text{cm}^2\,(R_2)(R_2)$$
$$= \frac{\sigma T^4}{\pi}\frac{a_1}{f_1^2}R_1^2\,R_2^2\,A_2\ \text{watts}. \qquad (8.5)$$

In our discussion concerning Figure 8.10 we were interested in transferring *total radiated power* from a reference radiation source located in the focal plane of the collimator on the left to the collimator on the right. Perhaps it should be emphasized that a collimator is basically a telescope and that a collimator is also basically a radiometer, except it is lacking a filter, a detector, and a radiation chopper.

Let us suppose that we now take the collimator shown on the right in Figure 8.10 and make it into a radiometer, i.e., we add a detector in the focal plane and in front of the detector we locate a radiation chopper. For the present, let us assume that no filter is to be used in front of the detector.

What we would like to do is to allow known power levels of radiation to fall on the collimator (now considered our radiometer) on the right and to observe the recorder deflection (measured in millivolts) that is produced. This procedure enables us to determine its calibration factor, or sensitivity factor, K, in watts/millivolt. In this case, we would like to make this determination not for total power but for known power levels in various wavelength intervals. In other words, we shall have to make an additional change in the experimental setup shown in Figure 8.10. In particular, we shall have to move the reference source as shown with the collimator on the left and where the

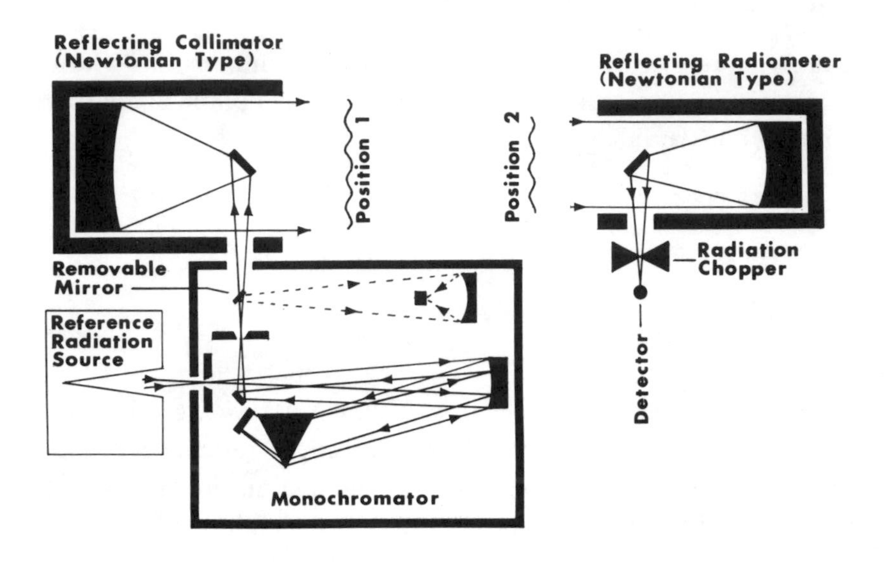

Figure 8.11: Experimental setup for radiometrically calibrating a reflecting radiometer.

source is shown we must locate at that point the *exit slit* of a spectrometer. At the *entrance slit* of the spectrometer we place the reference source. Figure 8.11 shows the experimental setup we would use.

8.6 Radiometric Calibration of a Thermopile

Next we consider the radiometric calibration of a thermopile. In particular, let us consider a thermopile manufactured by the Eppley Corporation. It will be recalled that a thermopile consists of two of more thermocouples connected in series. These

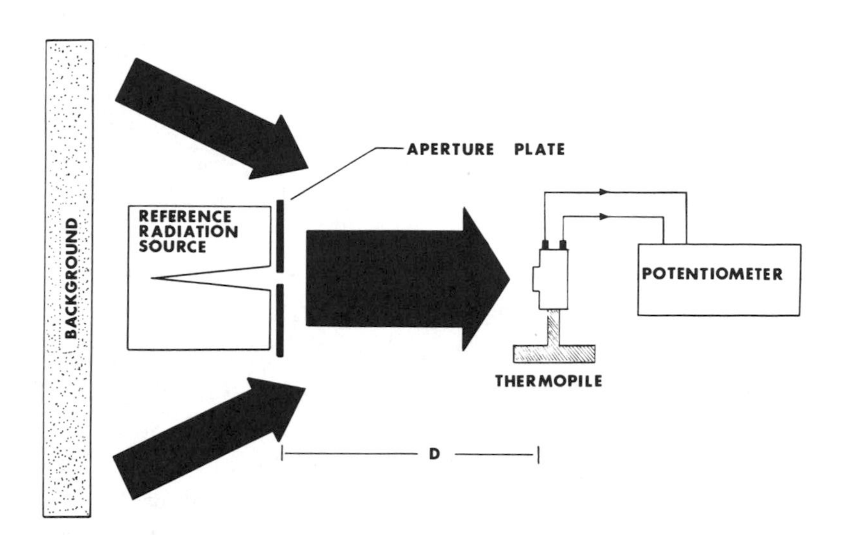

Figure 8.12: Experimental setup for radiometrically calibrating a thermopile.

are low impedance devices (a few ohms) and are nonselective detectors. The time constant is long (as far as detectors are concerned) so that one must use them with direct current applications. Figure 8.12 shows an experimental setup for carrying out our experiment. In performing this radiometric calibration, it is imperative that the background does not change during the process of calibration. Let

T_1 = temperature of the reference source;

a = area of opening in aperture plate located in front of the reference source;

A = area of thermopile receiver;

D = distance between reference source and the thermopile;

P_1 = total radiant power incident on thermopile due to the reference source:

$$P_1 = \frac{\sigma T_1^4}{\pi} a \frac{A}{D^2} \text{ watts.} \tag{8.6}$$

Let

P_0 = total radiant power incident on thermopile due to background radiation (watts);
P' = total power incident on thermopile due to all causes;
$P' = P_1 + P_0$.
This P' produces a certain reading on the potentiometer. Let M_1 = potentiometer reading (in millivolts).

We continue our experiment and raise the temperature of the reference source to some higher temperature, T_2, i.e., $T_2 > T_1$. Let
P_2 = total radiant power incident on the thermopile due to the reference source being operated at temperature T_2

$$P_2 = \frac{\sigma T_2^4}{\pi} a \frac{A}{D^2} \text{ watts.} \tag{8.7}$$

Let
P'' = total power incident on thermopile due to all causes;

$$P'' = P_2 + P_0. \tag{8.8}$$

This P'' produces a certain reading on the potentiometer. Let M_2 = potentiometer reading (in millivolts). Since $P'' > P'$, it follows that $M_2 > M_1$. To eliminate the contribution due to the background (and incidentally, we do not know what this contribution is), we calculate $(P'' - P')$:

$$\begin{aligned} P'' - P' &= (P_2 + P_0) - (P_1 + P_0) = P_2 - P_1 \\ &= \frac{\sigma T_2^4}{\pi} a \frac{A}{D^2} - \frac{\sigma T_1^4}{\pi} a \frac{A}{D^2} \\ &= \frac{\sigma}{\pi} a \frac{A}{D^2} (T_2^4 - T_1^4) \text{watts.} \end{aligned} \tag{8.9}$$

To arrive at a value for the sensitivity factor, K (watts per millivolt), we proceed as follows:

$$K = \frac{P'' - P'}{M_2 - M_1} = \frac{\sigma a A(T_2^4 - T_1^4)}{\pi D^2(M_2 - M_1)} \frac{\text{watts}}{\text{millivolt.}} \tag{8.10}$$

8.7 Choosing a Reference Radiation Source Temperature

In carrying out radiometric (or absolute) calibration, one should attempt to calibrate using a reference radiation source temperature that is as close as possible to the temperature of the radiating source to be spectroscopically or radiometrically studied. Since the highest temperature of commercially available reference radiation sources is about 1700°C, this could introduce a problem if one needs to calibrate at higher temperatures. Higher temperature reference sources are difficult to maintain and operate. One could, of course, calibrate the positive crater of a low-intensity carbon arc against one of these higher temperature reference radiation sources and use the carbon-arc-positive crater as a secondary standard.

From our discussion, one sees the general approach to carrying out radiometric calibrations, namely, choose two different temperatures at which to operate the reference source and for each of these two temperatures record the deflection produced by the spectrometer (or radiometer) and detector on the recorder. Then calculate the power falling upon the detector at each of these two temperatures for a particular spectral slit width (or spectral bandpass) and subtract these two values.

It should be emphasized again that this subtraction process eliminates the contribution due to the background. The hatched area shown in Figure 8.13 indicates what one obtains in this process of subtraction. One also subtracts the recorder deflections to get the recorder deflection corresponding to the hatched area shown in Figure 8.13.

For various reasons, one might not have access to a reference radiation source. Figure 8.14 is a sketch of a homemade source that can be operated at the temperature of ice and water and at the temperature of boiling water (both temperatures, of course, being converted to Kelvin). It is quite useful in undergraduate teaching laboratories and is easily constructed.

Radiometric calibration can be quite open-ended. For example, once one has radiometrically calibrated a spectrometer, such a system could be easily used to measure the spectral radiant emittance across the face of the Sun as a function of wavelength. Another possibility would be to measure the spectral radiant emittance of the positive crater of a low-intensity carbon arc. Most undergraduate laboratories are probably not equipped with enough d.c. power to operate a high-intensity carbon arc.

Another suggestion that is less demanding in the way of electrical power requirements is the measurement of the spectral radiant emittance of a globar infrared source and a Nernst filament infrared source as a function of wavelength. If one has access to a thermopile, the solar constant could be measured.

8.8 Use of a Collimator in the Laboratory

Collimators are extremely useful instruments to have available in the laboratory. You have seen some uses of collimators in this chapter. Two other uses for them are the following:

- for focusing telescopes and radiometers for use in parallel light, and

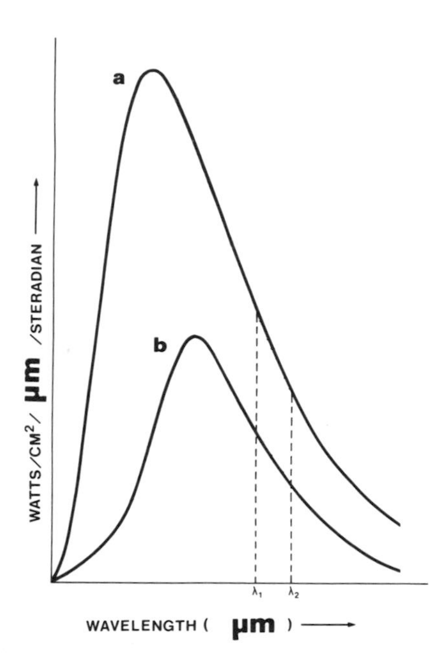

Figure 8.13: For a given spectral slit width, $(\lambda_2 - \lambda_1)$, the hatched area in the figure represents the difference in output of a reference source when operated at two different temperatures [curve (a) corresponds to a higher temperature than curve (b)].

- for measuring the field of view of a telescope and spectrometer system or a radiometer.

It is difficult to keep a large and heavy optical system pointed at a distant planet at night while the system is focused. However, it is simple and straightforward to direct toward the telescope, or radiometer, mirror a beam of parallel light from a collimator.

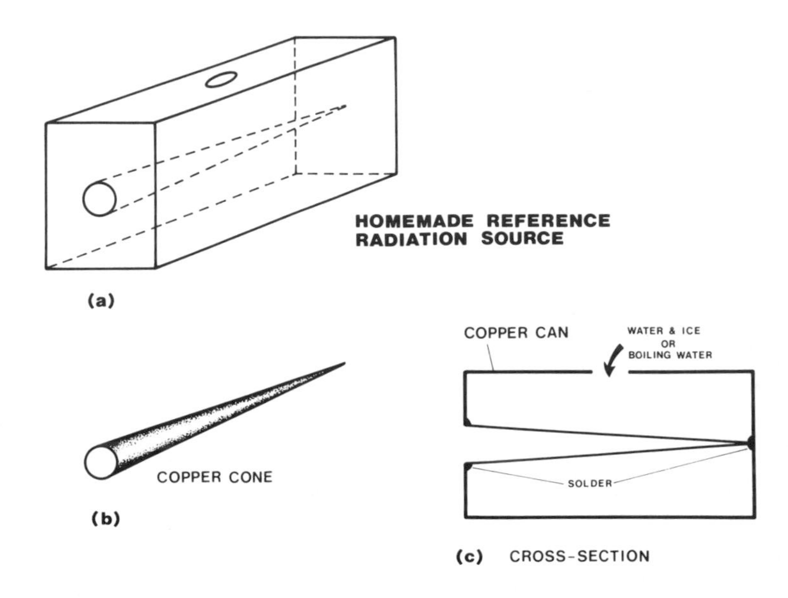

Figure 8.14: Homemade reference radiation source. Typical dimensions of the copper cone might be an opening of 1 1/2 inches and a length of 6 inches. By changing this ratio, one can change the radiant emissivity of the cone.

It is also easier to measure the field of view using a collimator. Referring again to Figure 8.10, suppose the collimator on the right is a radiometer and we want to measure its field of view. By rotating the radiometer about a vertical axis and measuring the output of the system as a function of the angle turned through, the azimuth scan can be obtained. A similar, but more difficult, procedure can be carried out to obtain the elevation scan.

References

1. E. K. Plyler, A. Danti, L. R. Blaine, and E. D. Tidwell. *Vibration-Rotation Structure in Absorption Bands for the Calibration of Spectrometers from 2 to 16 microns.* Natl. Bur. Stand. (U.S.) Monogr. 16 (1960)

2. Downie, Magoon, Purcell, and Crawford. "The Calibration of Infrared Prism Spectrometers." *J. Opt. Soc. Am.* **43**, 941- 951 (1953).

3. S. George and M. Voelker. "Calibration of prismatic spectra by means of Edser-Butler bands." *Am. J. Phys.* **36**, 1280 (1978).

4. X. Pivovansky and X. Nagel. *Tables of Blackbody Radiation Functions* (Macmillan, 1961).

5. R. G. Walker. *Tables of the Blackbody Radiation Function for Wavenumber Calculation.* Optical Phys. Lab. Project 4904 Rep. No. AFCRL-62-877 (Air Force Cambridge Research Laboratory, Bedford, MA, September, 1962).

Appendix

Some Useful Tables

TABLE 1

Energy Definitions and Relationships

1 erg = 1 dyne centimeter

1 joule = 1 Newton meter

1 joule = 10^7 ergs

1 joule = 0.7376 ft lb

1 ft lb = 1.356 joules

1 British thermal unit (BTU) = 252 calories

1 calorie = 4.186 joules (mechanical equivalent of heat)

1 electronvolt (eV) = 1.60×10^{-19} joule

1 kT = energy = 1/40 eV at room temperature

Energy = (mass) (speed of light in a vacuum)^2

Kinetic energy = $\frac{1}{2}mv^2$

Gravitational potential energy = mgh

Elastic potential energy = $\frac{1}{2}Kx^2$

Electrostatic potential energy = $\frac{1}{4\pi\epsilon_0}\frac{q_1q_2}{r}$

Electromagnetic radiation of wavelength about 12,000 Å has an energy of 1 electronvolt (eV)

Electromagnetic radiation of wavelength about 5460 Å (green light) has an energy of 2 electronvolts (eV)

Rest energy of the electron = m_0c^2 = 511,000 eV = 0.511 MeV

Binding energy of the electrons in isolated atoms (i.e., the ionization energy) varies from a few eV up to about 100,000 eV

Work function = binding energy for the most loosely bound electrons in a metal = a few eV

1 joule/second = 1 watt

1 kilowatt hour = 3.6×10^6 joules

1 volt = 1 joule/coulomb

Charge (e) on the electron = 1.60×10^{-19} coulomb

Avogadro's number = 6.02×10^{23} molecules/gram mole

Loschmidt's number = 2.69×10^{19} molecules/cm^3

1 horsepower (hp) = 550 $\frac{\text{ft lb}}{\text{second}}$

1 horsepower = 746 watts

TABLE 2

Some Physical Constants Involved in Radiation Exchange

c_1 = first radiation constant = $2\pi hc^2$ = 3.7413×10^{-5} erg cm^2 sec^{-1}

c_2 = second radiation constant = $\dfrac{hc}{k}$ = 1.4388 cm(K)

h = Planck's constant = 6.626×10^{-34} joule second

c = speed of light in a vacuum = 2.9979×10^{10} cm sec^{-1}

k = Boltzmann's constant = 1.381×10^{-23} joule (K^{-1})

σ = Stefan–Boltzmann constant = 5.6699×10^{-12} watt(cm^{-2}) (K^{-4})

Wein Displacement Law: $\lambda_{max} T$ = 2897.9 μm (K)

Planck's Radiation Law: 3 db points

$\lambda_{short} T = 5.1 \times 10^3$ μm (K)

$\lambda_{long} T = 1.8 \times 10^3$ μm (K)

Solar constant = 1400 watts/meter^2

TABLE 3

Wavelength Units

Spectral Region	Name	Definition
Gamma rays	X-unit = 10^{-3} Å	10^{-11}cm
	Angstrom (Å)	10^{-8}cm
X-rays	X-unit = 10^{-3} Å	10^{-11}cm
	Angstrom (Å)	10^{-8}cm
Ultraviolet radiation	Nanometer (nm)	10^{-9}m
	Angstrom (Å)	10^{-8}cm
	[10 angstroms = 1 nanometer]	
Visible radiation	Nanometer (nm)	10^{-9}m
	Angstrom (Å)	10^{-8}cm
Infrared radiation	Micron (μ)	10^{-4}cm
	Micrometer (μm)	10^{-6}m
	[1 micron = 1 micrometer]	
	[1 micron = 1,000 nm = 10,000 Å]	
Microwave region	Millimeters (mm)	
	Centimeters (cm)	
Radio region	Centimeters (cm)	
	Meters (m)	

TABLE 4

Radiometric Terminology, Symbols, Definitions, and Units
(Note: Not all of these terms have been used in this book.)

Radiometric Terminology	Symbol	Definition	Units
Radiant Power	P	Rate of transfer of energy	watt
Spectral Radiant Power	P_λ	Rate of transfer of energy per unit interval of wavelength	watt/μm
Radiant Intensity	J	Power per unit solid angle radiated by source	watt/steradian
Spectral Radiant Intensity	J_λ	Power per unit solid angle per unit interval of wavelength radiated by source	watt/steradian/μm
Radiance	N	Power per unit solid angle per unit area of source radiated by source	watt/steradian/cm^2
Spectral Radiance	N_λ	Power per unit solid angle per unit area of source per unit interval of wavelength radiated by source	watt/steradian/cm^2/μm
Irradiance	H	Power per unit area incident upon a surface	watt/cm^2

Spectral Irradiance	H_λ	Power per unit area per unit interval of wavelength incident upon a surface	watt/cm^2/μm
Radiant Energy	U	Energy radiated by a source	joule
Spectral Radiant Energy	U_λ	Energy radiated per unit interval of wavelength by source	joule/μm
Radiant Emittance	W	Power per unit area radiated from a surface	watt/cm^2
Spectral Radiant Emittance	W_λ	Power per unit area per unit interval of wavelength radiated by source	watt/cm^2/μm
Radiant Emissivity	ϵ	Ratio of ''emitted'' radiant power to that from a blackbody at the same temperature	
Radiant Absorptance	α	Ratio of ''absorbed'' radiant power to incident radiant power	
Radiant Reflectance	ρ	Ratio of ''reflected'' radiant power to incident radiant power	
Radiant Transmittance	τ	Ratio of ''transmitted'' radiant power to incident radiant power	

Radiometry and photometry involve the measurement, characterization, and interpretation of electromagnetic radiation. Radiometry is concerned with all wavelengths while photometry is concerned with the visible.

TABLE 5

Electromagnetic Spectrum, Energies Involved, and Physical Processes Involved

Physical Processes Involved	Energies Involved	Wavelengths Involved	Electromagnetic Spectrum	Frequencies Involved
Nuclear transitions			GAMMA RAYS	
	1.24×10^5 eV	0.10 Å	- - - - - - - - - - -	3×10^{19} Hz
			X-RAYS	
	12.4 eV	1000 Å	- - - - - - - - - - -	3×10^{15} Hz
			ULTRAVIOLET	
Atomic electronic transitions	3.1 eV	4000 Å (0.4 μm)	- - - - - - - - - - -	0.75×10^{15} Hz
			VISIBLE SPECTRUM	
	1.8 eV	7000 Å (0.7 μm)	- - - - - - - - - - -	0.43×10^{15} Hz
			NEAR INFRARED	
	0.83 eV	1.5 μm	- - - - - - - - - - -	2×10^{14} Hz
			INTERMEDIATE INFRARED	
Molecular vibration transitions	0.12 eV	10 μm	- - - - - - - - - - -	3×10^{13} Hz
			FAR INFRARED	
Molecular rotation transitions	0.0012 eV	1000 μm (1 mm)	- - - - - - - - - - -	3×10^{11} Hz
			MICROWAVES	
	0.000012 eV	100 mm (10 cm)	- - - - - - - - - - -	3×10^9 Hz
Accelerated nonbound electrons			RADIO	

The wavelength regions have not been drawn to scale.

TABLE 6

Some Infrared Detectors

Physical Mechanism of Transduction (or Detection)	Ease of Operation	Operating Temperature	Impedance (Ohms)	Time Constant (Seconds)	Spectral Response
Thermal					
1. Thermocouple	Straightforward	Room temperature (300 K)	5–10	10^{-3}–10^{-2}	Nonselective
2. Thermopile	Straightforward	Room temperature	10–100	10^{-2}	Nonselective
3. Pneumatic	Straightforward	Room temperature	- - - - -	10^{-2}	Nonselective
4. Thermistor bolometer	Straightforward	Room temperature	10^{6}	10^{-3}–10^{-2}	Nonselective
5. Thermal imaging devices	Difficult	Room temperature	- - - - -	Seconds	Nonselective
Quantum					
A. *PHOTOCONDUCTIVE*					
1. Lead sulfide	Straightforward	Room temperature and dry ice (194 K)	10^{6}–10^{8}	10^{-4}–10^{-3}	(1–3.5) μm
2. Lead selenide	Straightforward	Dry ice and liquid nitrogen (77 K)	10^{5}–10^{7}	10^{-5}–10^{-4}	(1–6) μm
3. Lead telluride	Straightforward	Dry ice and liquid nitrogen	10^{8}	10^{-5}–10^{-4}	(1–5.5) μm
4. Gold-doped geramium	Difficult	Liquid helium (4 K)	10^{6}–4×10^{7}	10^{-5}–10^{-3}	(1–9) μm
B. *PHOTOVOLTAIC*					
1. Indium antimonide	Straightforward	Room temperature, dry ice, liquid nitrogen	10^{2}–10^{3}	10^{-6}	(1–5.5) μm

TABLE 7

Prefixes for Powers of 10 When Using Large or Small Numbers

	Prefix	Symbol	Power of 10		
Greek Prefixes	exa	E	10^{18} =	1,000,000,000,000,000,000	Greater than one
	peta	P	10^{15} =	1,000,000,000,000,000	
	tera	T	10^{12} =	1,000,000,000,000	
	giga	G	10^{9} =	1,000,000,000	
	mega	M	10^{6} =	1,000,000	
	kilo	k	10^{3} =	1,000	
	hecto	h	10^{2} =	100	
	deca	da	10^{1} =	10	
Latin Prefixes	deci	d	10^{-1} =	0.1	Less than one
	centi	c	10^{-2} =	0.01	
	milli	m	10^{-3} =	0.001	
	micro	μ	10^{-6} =	0.000001	
	nano	n	10^{-9} =	0.000000001	
	pico	p	10^{-12} =	0.000000000001	
	femto	f	10^{-15} =	0.000000000000001	
	atto	a	10^{-18} =	0.000000000000000001	

Index

R

S

T

U

V

W

X

Y